Student Study Problems Supplement

D0220322

FUNDAMENTALS OF THERMODYNAMICS

Sixth Edition

RICHARD E. SONNTAG
CLAUS BORGNAKKE
University of Michigan

GORDON J. VAN WYLEN
Emeritus
Hope College

WILEY

John Wiley & Sons, Inc.

To order books or for customer service call 1-800-CALL-WILEY (225-5945).

Copyright © 2005 John Wiley & Sons, Inc. All rights reserved.

No part of this publication may be reproduced, stored in a retrieval system or transmitted in any form or by any means, electronic, mechanical, photocopying, recording, scanning or otherwise, except as permitted under Sections 107 or 108 of the 1976 United States Copyright Act, without either the prior written permission of the Publisher, or authorization through payment of the appropriate per-copy fee to the Copyright Clearance Center, 222 Rosewood Drive, Danvers, MA 01923, (978) 750-8400, fax (978) 750-4470. Requests to the Publisher for permission should be addressed to the Permissions Department, John Wiley & Sons, Inc., 111 River Street, Hoboken, NJ 07030, (201) 748-6011, fax (201) 748-6008, E-Mail: PERMREQ@WILEY.COM.

ISBN 0-471-69648-X

Printed in the United States of America

10 9 8 7 6 5 4 3

Printed and bound by Hamilton Printing Company

CONTENTS

CHAPTER 2
STUDY PROBLEMS
INTRODUCTION

- **The thermodynamic system and control volume**
- **Macroscopic versus microscopic point of view**
- **Properties and state of a substance**
- **Processes and cycles**
- **Units**
- **Energy**
- **Specific volume and density**
- **Pressure**
- **Temperature**

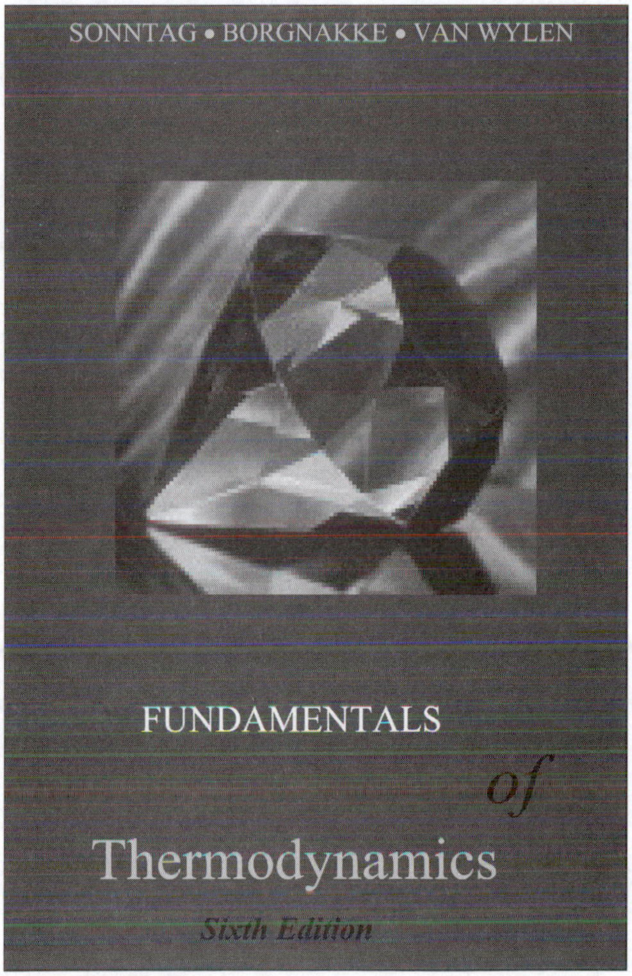

2.1 A hydraulic lift

A hydraulic cylinder has a diameter of 7.5 cm. What should the fluid pressure inside be to create a force of 6000 N on the rod if we neglect the outside pressure?

Solution:

From the definition of pressure as force per unit area

$$F = P\,A \qquad \text{or} \qquad P = \frac{F}{A}$$

The cross sectional area is

$$A = \pi\, r^2 = \pi\, D^2/4 = \frac{\pi}{4}\; 0.075^2\ \text{m}^2 = 0.004418\ \text{m}^2$$

so the pressure becomes

$$P = \frac{F}{A} = \frac{6000\ \text{N}}{0.004418\ \text{m}^2} = 1\,358\,080\ \text{Pa} = \mathbf{1358\ kPa}$$

Remark:

The outside atmospheric pressure of 101.325 kPa should be subtracted when calculating the force and thus be added to the above pressure, or you may think of the above pressure as the gauge pressure.

2.2 Static pressure and buoyancy force on a submerged vessel

A research vessel is built as a small submarine capable of going down to depth of 2000 m in the ocean. Two pressurized air tanks are used to blow air into the water ballast tanks to control the buoyancy, so when air enters the ballast tanks pushing water out the vessel will rise up. For safety we need these air tanks to have a pressure 100% higher than the water pressure. To what pressure should these air tanks be charged? After pushing 0.5 m^3 water out how large a force up does that add?

Solution:

The total water pressure is

$$P = P_{top} + \Delta P = P_{top} + \rho g h$$

Density of pure water from Table A.4 is $\rho = 997\ kg/m^3$ which we use as an approximation for salt water and it does not depend on pressure.

$$P = 101\ kPa + 997\ kg/m^3 \times 9.81\ m/s^2 \times 2000\ m \times \frac{1\ kPa}{1000\ Pa} = 19\ 662\ kPa$$

$$P_{tank} = 2 \times P_{water} = \mathbf{39\ 324\ kPa = 39.3\ MPa}$$

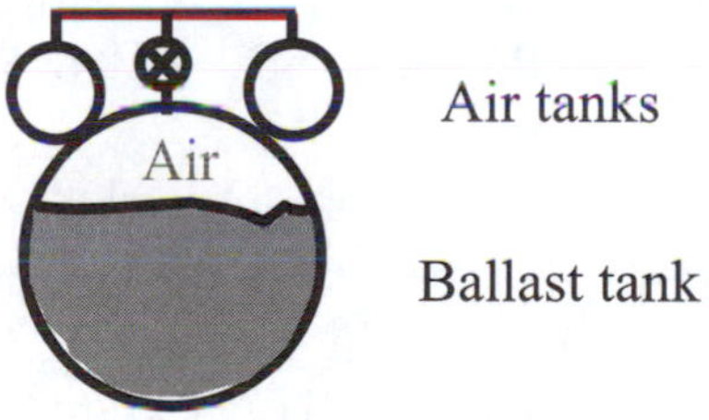

The air in the tanks now occupies a larger volume having the same mass and so the increased force up is equal to the gravitational force on the displaced water.

$$F = mg = \rho V g = 997\ kg/m^3 \times 0.5\ m^3 \times 9.81\ m/s^2 = \mathbf{4890\ N}$$

2.3 A manometer with glycerine

A manometer using glycerine as fluid is used to measure the pressure in a box where air flows. The hole in the box is 1 m higher than the glycerine surface in the manometer showing a pressure difference of positive 0.25 m glycerine. Assuming an atmospheric pressure of 101 kPa find the gauge pressure and absolute pressure at the hole in the box. Repeat the answers if the box flows liquid water instead of air.

Solution:

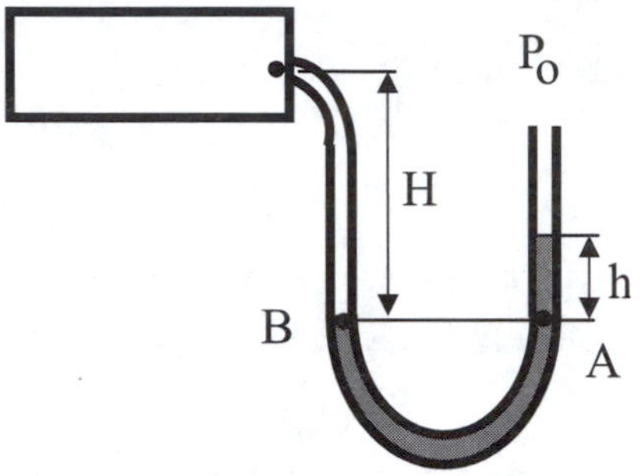

The density of glycerine from Table A.4 is $\rho = 1260$ kg/m^3.
Relate the pressure P_C to P_B which equals P_A (liquid below A and B is in equilibrium).

The top surface above A has atmospheric pressure so

$$P_A = P_{top} + \Delta P = P_o + \rho g H$$

$$= 101 \text{ kPa} + 1260 \text{ kg/m}^3 \times 9.81 \text{ m/s}^2 \times 0.25 \text{ m} \times \frac{1 \text{ kPa}}{1000 \text{ Pa}} = 104.09 \text{ kPa}$$

If we neglect the change in P from B to C (air has low density) we have

$$P_C = P_B = P_A = \mathbf{104.09\ kPa}\ ; \qquad \Delta P_C = P_C - P_o = \mathbf{3.09\ kPa}$$

To check the assumption, air is close to 100 kPa so from Fig. 2.7 or Table A.5 we have $\rho = 1$ kg/m^3 (or 1.17). Then the pressure difference between B and C is

$$\Delta P = \rho g H = 1 \text{ kg/m}^3 \times 9.81 \text{ m/s}^2 \times 1 \text{ m} \times \frac{1 \text{ kPa}}{1000 \text{ Pa}} = 0.0098 \text{ kPa}$$

which is insignificant, validating our assumption. If we have liquid water flowing assume the fluid between B and C is liquid water with a density from Table A.4 as $\rho = 997$ kg/m^3. In that case we have

$$\Delta P = P_B - P_C = \rho g H$$

$$= 997 \text{ kg/m}^3 \times 9.81 \text{ m/s}^2 \times 1 \text{ m} \times 1 \text{ kPa} / 1000 \text{ Pa} = 9.78 \text{ kPa}$$

so then

$$P_C = P_B - \Delta P = P_A - \Delta P = 104.09 - 9.78 = \mathbf{94.31\ kPa}$$

with a gauge pressure of

$$\Delta P_C = P_C - P_o = 94.31 - 101 = \mathbf{-6.69\ kPa}$$

and we notice it is a vacuum.

2.4 A dual acting piston-cylinder arrangement

Two cylinders are connected with a 15 kg piston as shown. Cylinder A has light oil in it with a height of 1 m to the piston surface of area 0.1 m^2. Cylinder B has air with a small height and an area of 0.9 m^2. A pump delivers oil at the bottom of cylinder A with a pressure P. Find P so the pressure in B is 250 kPa assuming the atmosphere is at 101 kPa.

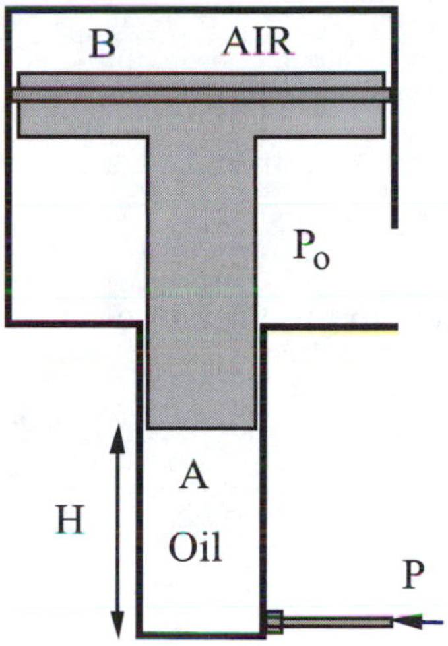

Solution:

Assume the piston is at rest so force balance:

$$F = m_p a = 0 = F\uparrow - F\downarrow$$

$$F\downarrow = P_B A_B + m_p g$$

$$F\uparrow = P_0(A_B - A_A) + P_{A\,top} A_A$$

Since the oil has some height with a density for light oil from Table A.4, $\rho = 910\ kg/m^3$, we can relate the pressures at the bottom and top as

$$P = P_{A\,bot} = P_{A\,top} + \rho g H \qquad \Rightarrow \qquad P_{A\,top} = P - \rho g H$$

substitute this into the force balance equation

$$P_B A_B + m_p g = P_0(A_B - A_A) + (P - \rho g H) A_A$$

and solve for P

$$P = \rho g H + P_B \frac{A_B}{A_A} + \frac{m A_p g}{A_A} - P_0 \frac{A_B - A_A}{A_A}$$

$$= 910 \frac{kg}{m^3} \times 9.81 \frac{m}{s^2} \times 1\ m + 250\ kPa \frac{0.9\ m^2}{0.1\ m^2} + \frac{15\ kg \times 9.81\ m/s^2}{0.1\ \ m^2}$$

$$-\ 101\ kPa \frac{0.9 - 0.1}{0.1} \frac{m^2}{m^2}$$

$$= 8927.1\ Pa + 2250\ kPa + 1471.5\ Pa - 808\ kPa$$

$$= 8.927\ kPa + 2250\ kPa + 1.472\ kPa - 808\ kPa$$

$$= \mathbf{1452.4\ kPa}$$

Notice how the influence from the pressure difference in the oil and the piston mass is very small compared with the atmospheric and cylinder B pressures.

2.1E A hydraulic lift

A hydraulic cylinder has a diameter of 3 in. What should the fluid pressure inside be to create a force of 1500 lbf on the rod if we neglect the outside pressure?

Solution:

From the definition of pressure as force per unit area

$$F = P\,A \qquad \text{or} \qquad P = \frac{F}{A}$$

The cross sectional area is

$$A = \pi\, r^2 = \pi\, D^2/4 = \frac{\pi}{4}\, 3^2\ \text{in}^2 = 7.0686\ \text{in}^2$$

so the pressure becomes

$$P = \frac{F}{A} = \frac{1500\ \text{lbf}}{7.0686\ \text{in}^2} = \mathbf{212\ psi = 212\ \frac{lbf}{in^2}}$$

Remark:

The outside atmospheric pressure of 14.7 psia should be subtracted when calculating the force and thus be added to the above pressure, or you may think of the above pressure as the gauge pressure.

CHAPTER 3
STUDY PROBLEMS
PROPERTIES OF A PURE SUBSTANCE

- **The vapor-liquid-solid phase equilibrium in a pure substance**
- **Tables of thermodynamic properties**
- **The ideal gas law**
- **Compressibility factor**
- **Computerized tables**

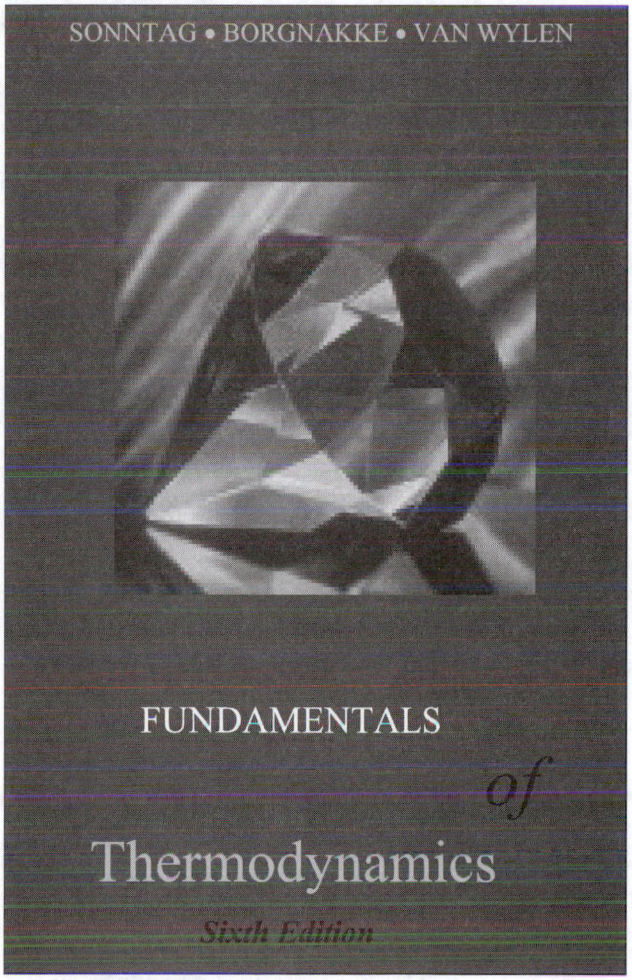

3.1 The vaporization line and the critical point

What is the phase at room pressure, 101 kPa and temperature, 20°C, for the following substances: Oxygen, carbon dioxide, water and R-134a.

Solution:

To answer the question we need to know something about the location of the phase boundaries. For the critical point we have information from Table 3.1 or A.2.

Room	$T = 293$ K	$P = 101$ kPa
Oxygen	$T_c = 155$ K	$P_c = 5040$ kPa
Carbon dioxide	$T_c = 304$ K	$P_c = 7380$ kPa
Water	$T_c = 647$ K	$P_c = 22090$ kPa
R-134a	$T_c = 374$ K	$P_c = 4064$ kPa

From this information alone we can notice:

Oxygen has $T >> T_c$ and $P << P_c$ so **superheated vapor**, recall Fig. 3.5

Carbon dioxide has $T \cong T_c$ and $P << P_c$ so **superheated vapor**, recall Fig. 3.5, and the actual phase diagram in Fig.3.6 could be used.

Water has lower T and much lower P than the critical point so we need to know how much lower the saturation pressure is for the lower T as shown in Fig.3.7. From the vaporization line we estimate $P_{sat} = 3$ kPa, the higher P means a **compressed liquid** state. More accurately we could look in Table B.1.1 and find $P_{sat} = 2.3$ kPa.

R-134a has lower T and P than the critical point so we must look up in Table B.5.1. For 20°C we have a saturation pressure of about 573 kPa the lower room pressure means we have an expanded vapor (**superheated vapor**)

The states are shown relative to the vaporization line in the P-T diagram below.

The fusion line for water slants left for the others it slants right.

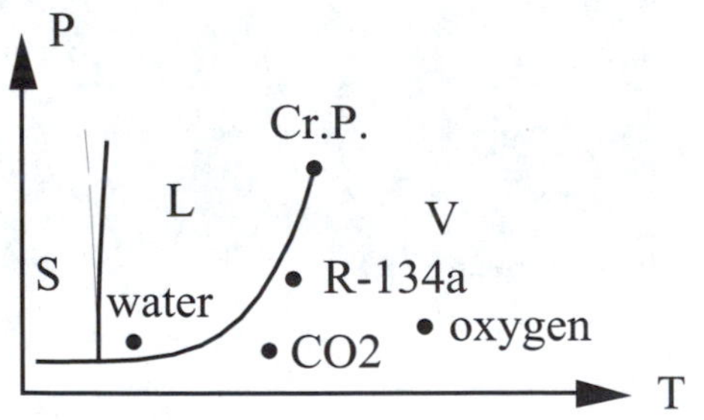

3.2 The waporization line for R-134a and methane

I. To design parts of the air-conditioning system for a car we would like to know the pressure we will see when R-134a is throttled to become two-phase L+V at –20°C. We also would like to know how hot is the liquid R-134a if we condense it at 800 kPa.

II. A storage facility for methane (main component of natural gas) will store it as a liquid at 150 K. A vent valve on the tank will open at a pressure of 2500 kPa in case the cooling system fails and the pressure starts to increase. What is the minimum normal storage pressure and what is the maximum temperature in the tank while we still have some liquid inside?

Solution:

I. A two phase state (L+V) is on the vaporization line with a P_{sat} given the temperature. From Table B.5.1 we see

$$P_{sat} = \mathbf{133.7\ kPa} \text{ at } -20^{o}\text{C}$$

We do not have a pressure entry table so we interpolate in the Table B.5.1 to get 800 kPa and obtain:

$$T_{sat} = 30 + (35 - 30)\,\frac{800 - 771.0}{887.6 - 771.0} = \mathbf{31.24^{o}C}$$

Notice the superheated vapor tables in B.5.2 at 800 kPa has 31.3°C for the first line.

II. We have a compressed liquid state when the pressure is higher than the saturation pressure. From table B.7.1 we see

$$P_{sat} = \mathbf{1040.5\ kPa} \text{ at } 150\text{ K}$$

so this constitutes the minimum needed P.
For a given P the maximum T, while we still have liquid, is the boiling temperature. From Table B.7.1 we interpolate to get

$$T_{sat} = 170 + (175 - 170)\,\frac{2500 - 2329.3}{2777.6 - 2329.3} = \mathbf{171.9\ K}$$

The diagrams are different for R-134a and methane

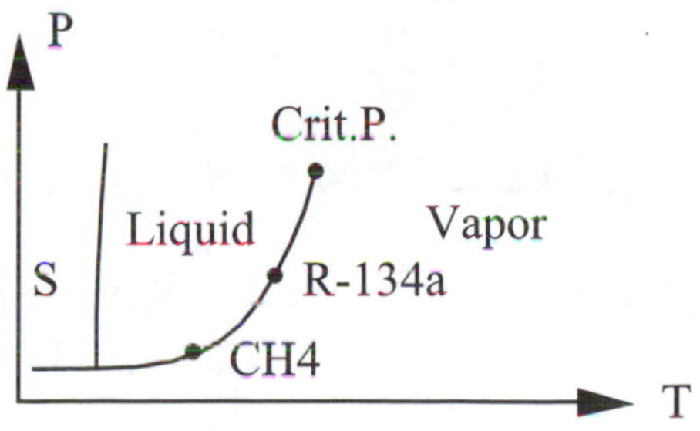

3.3 Water tables

Determine the missing property of P-v-T and x if applicable for the following states for water.

a) 140°C, 5000 kPa b) 200 kPa, 0.25 m^3/kg

Solution:

a) Enter Table B.1.1 with 140°C and we see a saturation pressure of 361.3 kPa, so at 5000 kPa we have a compressed liquid $P > P_{sat}$, see P-T diagram.

If we started in Table B.1.2 with 5000 kPa and then $T_{sat} = 263.99°C$, so we have compressed liquid (subcooled $T < T_{sat}$, see P-T diagram).

Proceed to Table B.1.4 subsection for 5000 kPa with the number after the 5000 kPa as T_{sat} (boiling T for that P), which here is 263.99°C. We get $\mathbf{v = 0.001077\ m^3/kg}$.

b) Enter Table B.1.2 with 200 kPa and notice

$$v_f = 0.001061 < v < v_g = 0.88573\ m^3/kg$$

we have a two-phase mixture of liquid and vapor at saturation **T = 120.2°C**. The quality is from Eq.3.10

$$x = (v - v_f)/v_{fg} = (0.25 - 0.001061) / 0.88467 = 0.2814$$

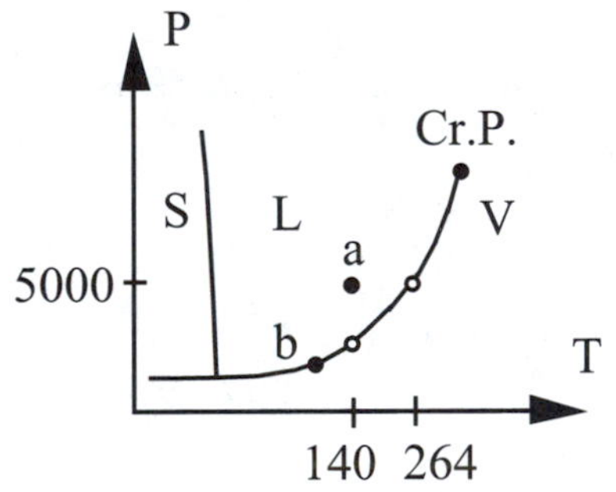

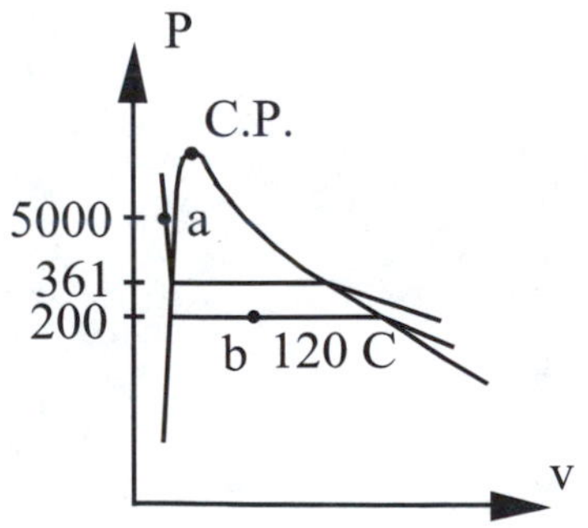

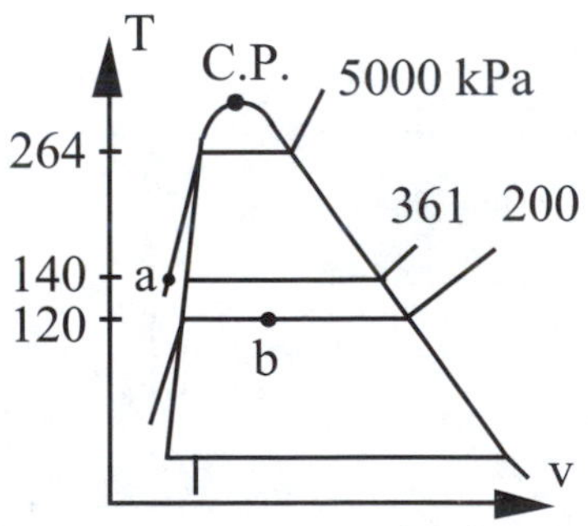

3.4 The R-22 and R-134a tables

Determine the missing property of P-v-T and x if applicable for the following states.

a) R-22: 30°C, 1000 kPa b) R-134a: 201 kPa, 0.1 m^3/kg

Solution:

a) For R-22 we look in table B.4.1 with T = 30°C and notice

$$P < P_{sat} = 1191.9 \text{ kPa}$$

so we have superheated vapor. Proceed to table B.4.2 subsection 1000 kPa (T_{sat} = 23.42°C). There we find the entry for v as: **v = 0.02460 m³/kg**

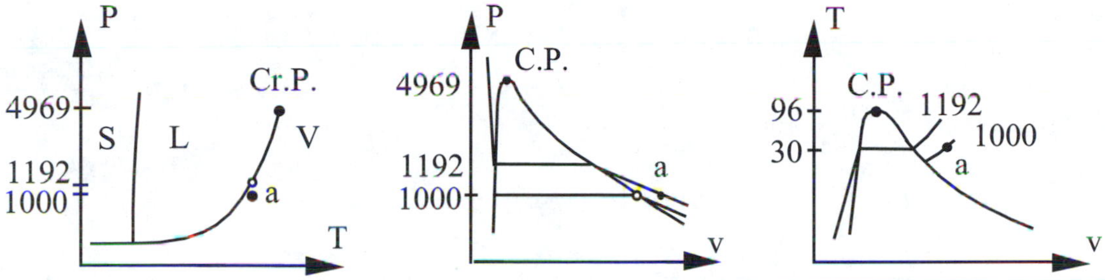

b) For R-134a we look in table B.5.1 with P = 201 kPa and notice that this is very close to the saturated vapor state listed at 201.7 kPa. See also in B.5.2 subsection 200 kPa where the first entry is the saturated vapor at 200 kPa. Interpolate between these two entries to get v_g at 201 kPa

$$v_g = 0.10002 - \frac{201 - 200}{201.7 - 200} (0.10002 - 0.09921) = 0.09954 \text{ m}^3\text{/kg}$$

We conclude the state is **superheated** (but very little) **vapor** at roughly **–10°C**.

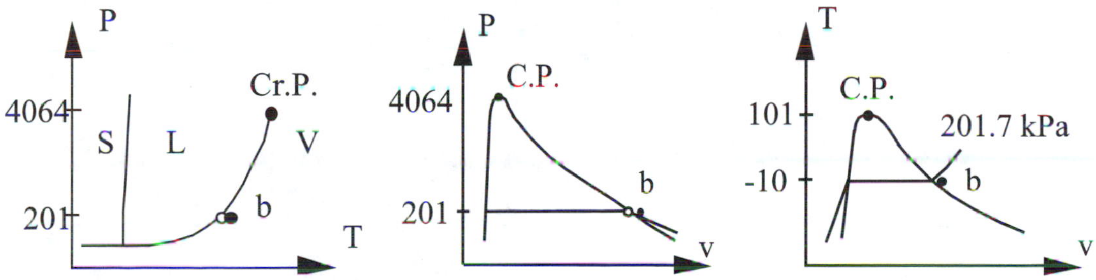

3.5 The methane table

A 3 m^3 rigid steel tank with liquified natural gas LNG (which we take as methane) is kept at 150 K and currently has a quality of 64%. There is a temperature sensor mounted on it and connected to a control and alarm system in addition to a pressure gauge. How much methane does the tank hold? At what temperature should the alarm sound if the tank should be at a pressure below 2000 kPa?

Solution:

The beginning state (a) is two-phase and we can find the properties in Table B.7.1:

$$v_a = v_f + x\, v_{fg} = 0.002794 + 0.64\ (0.05839) = 0.04016\ m^3/kg$$

The mass therefore becomes

$$m = V / v_a = 3 / 0.04016 = \mathbf{74.7\ kg}$$

When the tank and its content heats up we assume that the volume does not change so

Process: $v = \text{constant} = v_a$

At the final pressure of 2000 kPa we see in Table B.7.1 (or B.7.2) that $v_b = v_a > v_g$ so the state becomes superheated vapor. From interpolation in Table B.7.2 at 2000 kPa we get the temperature as

$$T = 175 + (200 - 175)\frac{0.04016 - 0.03504}{0.04463 - 0.03504} = \mathbf{188.3\ K}$$

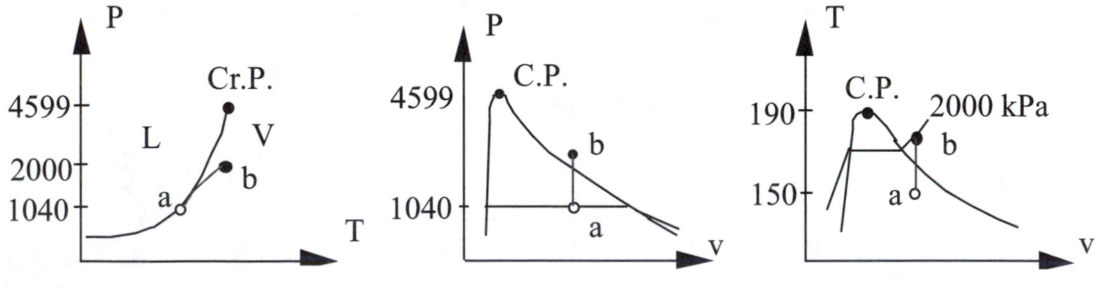

3.6 A hot-air balloon and ideal gas law

A hot-air balloon is inflated with hot air, 75°C, at approximately 101 kPa to a volume of 775 m^3. How much air mass is that? and how much air mass at 20°C, 101 kPa is displaced by the balloon volume?

Solution:

We assume the air is an ideal gas so we have from Eq.3.5

$$PV = mRT \;; \qquad R = 0.287 \text{ kJ/kg K from table A.5}$$

now

$$m = \frac{PV}{RT} = \frac{101 \times 775}{0.287 \times (273.15 + 75)} \frac{\text{kPa m}^3 \text{ kg K}}{\text{kJ K}} = \mathbf{783.4\ kg}$$

The mass of cold air displaced by the same volume is

$$m = \frac{PV}{RT} = \frac{101 \times 775}{0.287 \times (273.15 + 20)} \frac{\text{kPa m}^3 \text{ kg K}}{\text{kJ K}} = \mathbf{930.4\ kg}$$

So that size balloon with the hot air at 75°C can lift the difference (930 – 783) kg in the gravitational field, an effect called the buoyancy.

3.7 The compressibility chart

Determine the specific volume of nitrogen at 140 K and 3000 kPa using

a) The nitrogen table　　b) Ideal gas law

c) The compressibility chart

Solution:

a) For nitrogen we look in table B.6.2, since $T > T_c$ we have superheated vapor

$$\mathbf{v = 0.01038\ m^3/kg}$$

b) For the ideal gas law we have from table A.5: $R = 0.2968$ kJ/kg K

$$v = \frac{RT}{P} = \frac{0.2968 \times 140}{3000} = \mathbf{0.01385\ m^3/kg}$$

c) For the compressibility chart Fig. D.1 we need the critical constants
Table A.2 or B.6.2: $T_c = 126.2$ K, $P_c = 3390$ kPa

$$T_r = \frac{T}{T_c} = \frac{140}{126.2} = 1.109, \qquad P_r = \frac{P}{P_c} = \frac{3000}{3390} = 0.885$$

Now read from the chart Fig. D.1: $Z = 0.75$

$$v = \frac{ZRT}{P} = \frac{0.75 \times 0.2968 \times 140}{3000} = \mathbf{0.01039\ m^3/kg}$$

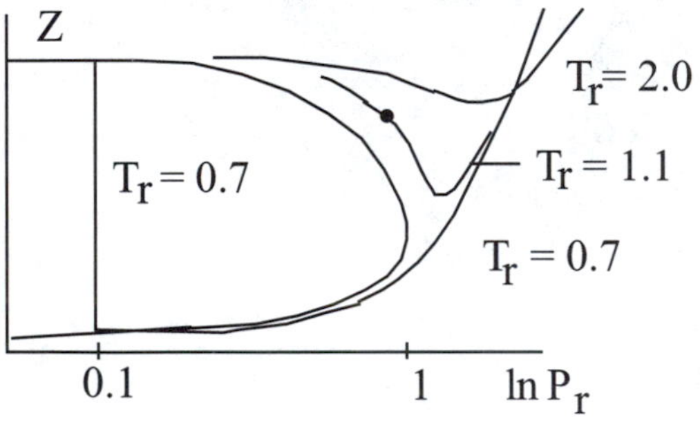

Notice how close the compressibility chart approximation is, within chart reading error, whereas the ideal gas approximation is 33% too high.

3.8 The van der Waal equation of state

Determine the specific volume of nitrogen at 140 K and 3000 kPa using van der Waal EOS.

Solution:

Table A.2 or B.6.2: $T_c = 126.2$ K, $P_c = 3390$ kPa

For van der Waal equation of state from Table D.1 we have

$$b = \frac{1}{8}\frac{RT_c}{P_c} = 0.125 \times \frac{0.2968 \times 126.2}{3390} = 0.001\ 381\ 1\ \text{m}^3/\text{kg},$$

$$a = 27\ b^2\ P_c = 27 \times (0.001\ 381\ 1)^2 \times 3390 = 0.174595\ \text{kPa}\ (\text{m}^3/\text{kg})^2$$

The equation is: $P = \frac{RT}{v - b} - \frac{a}{v^2}$ or $3000 = \frac{0.2968 \times 140}{v - 0.0013811} - \frac{0.174595}{v^2}$

so we have an implicit nonlinear equation for the specific volume. Use trial and error, start with $v = 0.01038$ m^3/kg (from the nitrogen table B.6.2) and solve for P

$$P = 4617.45 - 1620.45 = 2997 \text{ kPa} \quad \text{slightly too small}$$

Guess $v = 0.01037$ m^3/kg and solve again

$$P = 4622.59 - 1623.582 = 2999 \text{ kPa} \quad \text{close}$$

Guess $v = 0.010365$ m^3/kg and solve again

$$P = 4625.16 - 1625.15 = 3000 \text{ kPa} \quad \text{OK}$$

so the answer is **v = 0.010365 m^3/kg**

3.9 The computerized tables

Find the specific volume of ethylene at 25°C and 3000 kPa assuming ideal gas and using the computerized tables.

The gas constant from Table A.5 is R = 0.2964 kJ/kg K

$$v = \frac{RT}{P} = \frac{0.2964 \times (25 + 273.15)}{3000} = 0.02946 \text{ m}^3/\text{kg}$$

Using the software CATT2 we select the cryogenic substances, select ethylene and click on the small calculator button. Select case 1: (T, P) Input 25°C and 3 MPa. The following output is shown

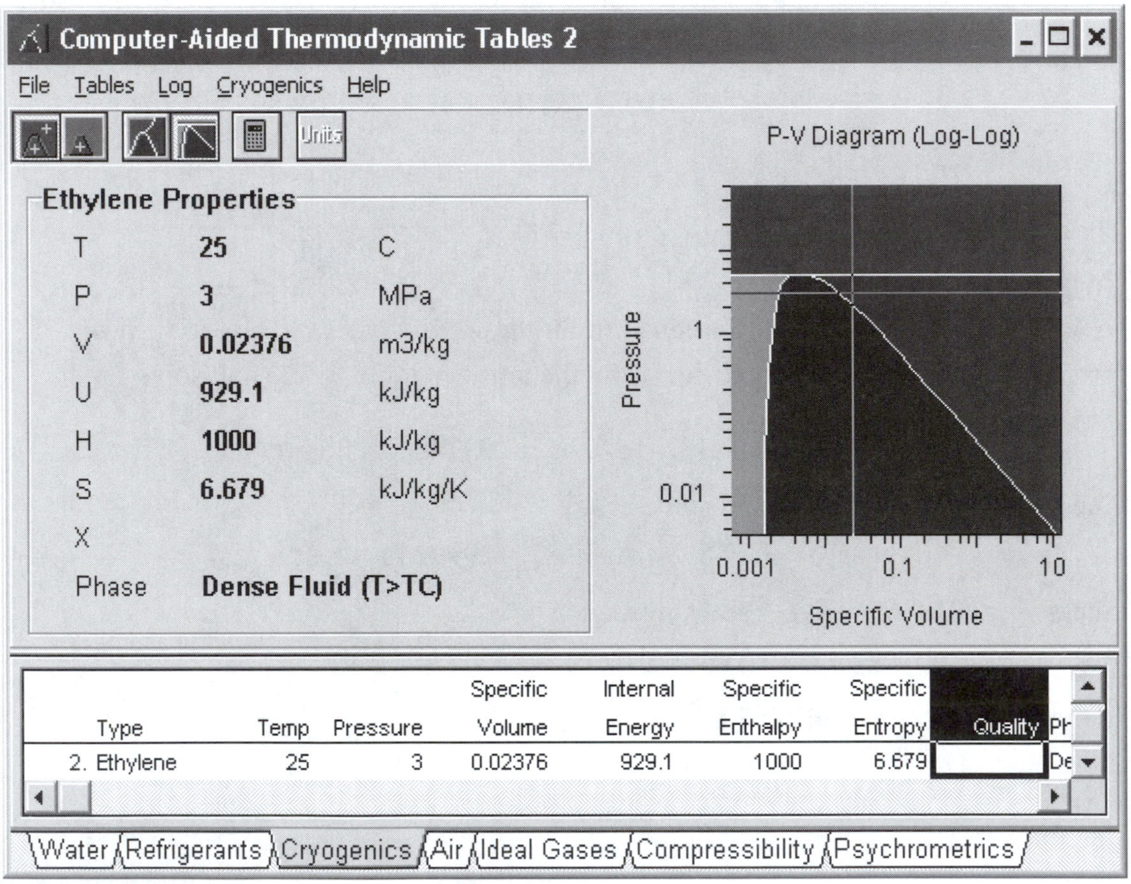

Here we see that

$$v = 0.02376 \text{ m}^3/\text{kg}$$

which is 20% smaller than for ideal gas. The difference is higher for a higher pressure.

3.3E Water tables

Determine the missing property of P-v-T and x if applicable for the following states for water.

a) 250 F, 500 psia b) 400 F, 0.5 ft^3/lbm

Solution:

a) Enter Table F.7.1 with 250 F and we see a saturation pressure of 29.823 psia, so at 500 psia we have a compressed liquid $P > P_{sat}$, see P-T diagram.
Proceed to Table F.7.3 subsection for 500 psia with the number after the 500 psia as T_{sat} (boiling T for that P), which here is 467.12 F. We get **v = 0.017 ft^3/lbm**.

b) Enter Table F.7.1 with 400 F and notice

$$v_f = 0.01864 < v < v_g = 1.866 \text{ ft}^3/\text{lbm}$$

we have a two-phase mixture of liquid and vapor at saturation **P = 347.08 psia**. The quality is from Eq.3.10

$$x = \frac{v - v_f}{v_{fg}} = \frac{0.5 - 0.01864}{1.8474} = 0.2606$$

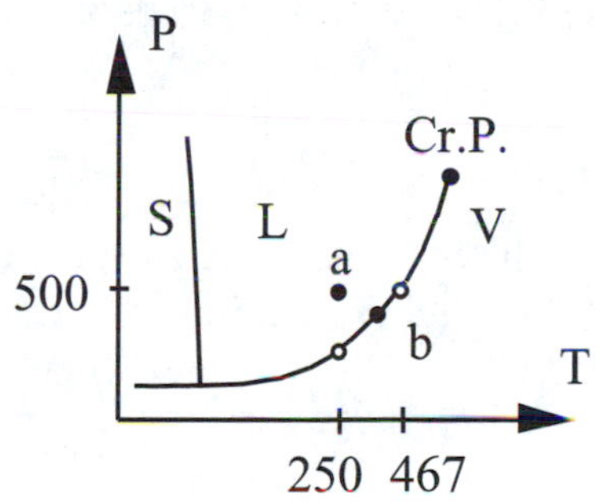

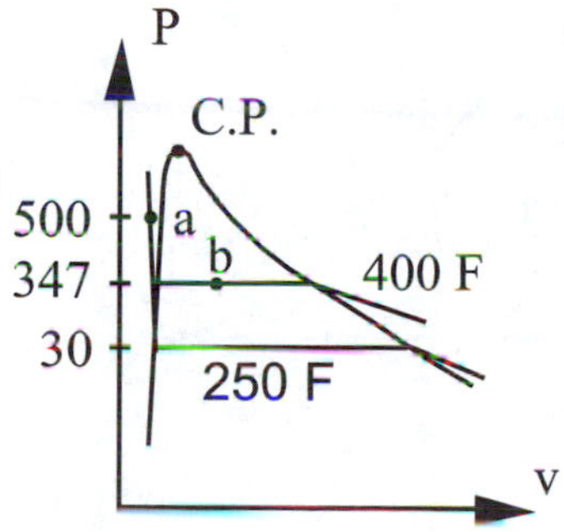

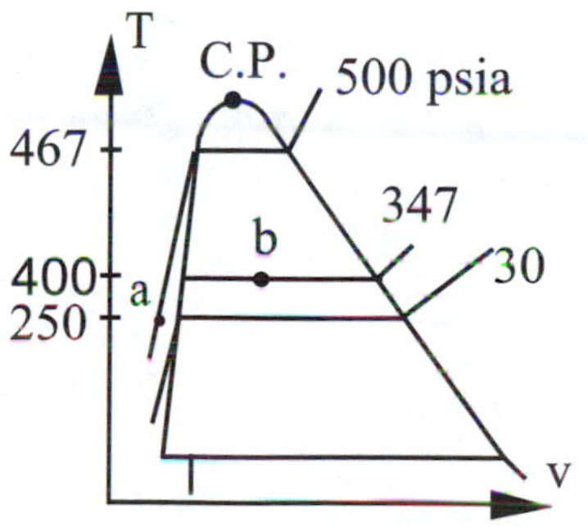

3.4E The R-22 and R-134a tables

Determine the missing property of P-v-T and x if applicable for the following states.

a) R-22: 80 F, 150 psia b) R-134a: 30 psia, 1.57 ft^3/lbm

Solution:

a) For R-22 we look in table F.9.1 with T = 80 F and notice

$$P < P_{sat} = 158.33 \text{ psia}$$

so we have superheated vapor. Proceed to table F.9.2 subsection 150 psia (T_{sat} = 76.38 F). There we find the entry for v as: **v = 0.3705 ft^3/lbm**

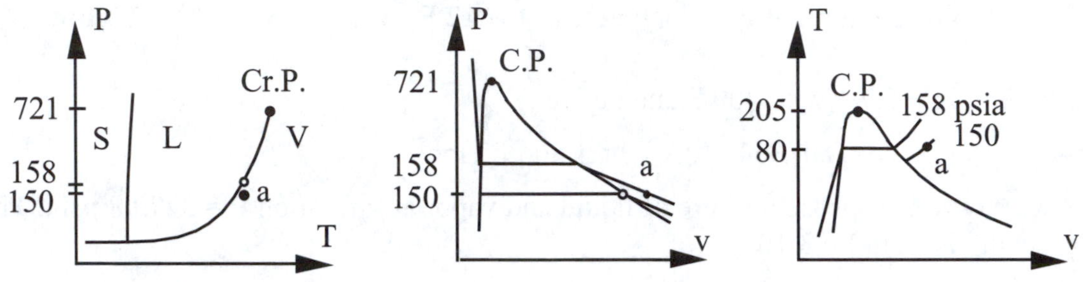

b) For R-134a we look in table F.10.1 with P = 30 psia and notice that this must be interpolated.

$$v_g = 1.7282 + (1.405 - 1.7282)\frac{30 - 26.787}{33.294 - 26.787} = 1.5686 \text{ ft}^3\text{/lbm}$$

since the actual v is larger we have superheated vapor. Proceed to Table F.10.2 in the 30 psia section (T_{sat} = 15.15 F) where we interpolate for the temperature

$$T = 15.15 + (20 - 15.15)\frac{1.57 - 1.5517}{1.5725 - 1.5517} = 19.42 \text{ F}$$

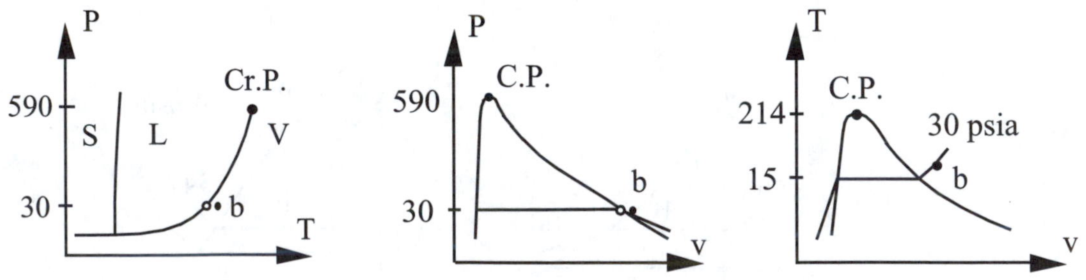

3.7E The compresibility chart

Determine the specific volume of R-134a at 280 F and 500 psia using

a) The R-134a table b) Ideal gas law

c) The compressibility chart

Solution:

a) For R-134a we look in table F.10.2, since $T > T_c$ we have superheated vapor

$$\mathbf{v = 0.1174\ ft^3/lbm}$$

b) For the ideal gas law we have from table F.4: R = 15.15 ft-lbf / lbm-R

$$v = \frac{RT}{P} = \frac{15.15 \times (280 + 459.7)}{500} \frac{ft \times lbf}{lbm \times R} \frac{R \times in^2}{lbf} \frac{ft^2}{144\ in^2} = \mathbf{0.1556\ ft^3/lbm}$$

c) For the compressibility chart Fig. D.1 we need the critical constants
Table F.1 or F.10.1: $T_c = 673.6$ R, $P_c = 589$ psia

$$T_r = \frac{T}{T_c} = \frac{280 + 459.7}{673.6} = 1.098, \qquad P_r = \frac{P}{P_c} = \frac{500}{589} = 0.849$$

Now read from the chart Fig. D.1: Z = 0.74

$$v = \frac{ZRT}{P} = Z\ v_{ideal\ gas} = 0.74 \times 0.1556\ ft^3/lbm = \mathbf{0.1152\ ft^3/lbm}$$

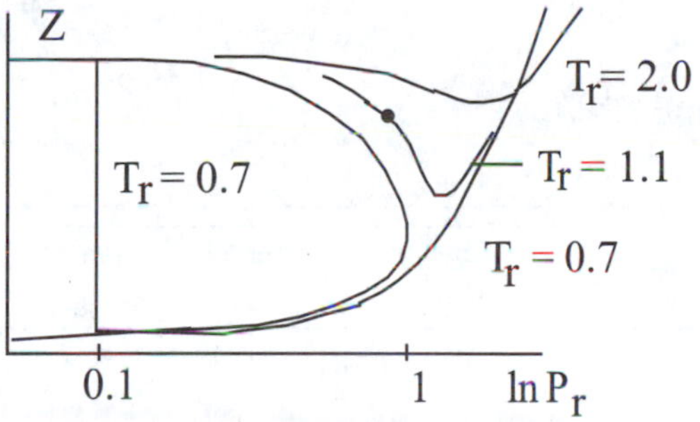

Notice how close the compressibility chart approximation is with 2% too low whereas the ideal gas approximation is nearly 33% too high.

3.9E The computerized tables

Find the specific volume of ethylene at 77 F and 400 psia assuming ideal gas and using the computerized tables.

The gas constant from Table F.4 is R = 55.07 ft lbf / lbm R

$$v = \frac{RT}{P} = \frac{55.07 \times (77 + 459.7)}{400} \frac{\text{ft} \times \text{lbf}}{\text{lbm} \times \text{R}} \frac{\text{R} \times \text{in}^2}{\text{lbf}} \frac{\text{ft}^2}{144\ \text{in}^2} = \mathbf{0.513\ ft^3/lbm}$$

Using the software CATT2 we select the cryogenic substances, select ethylene and click on the unit button. Click on the small calculator button.

Select case 1: (T, P) Input 77 F and 400 psi. The following output is shown

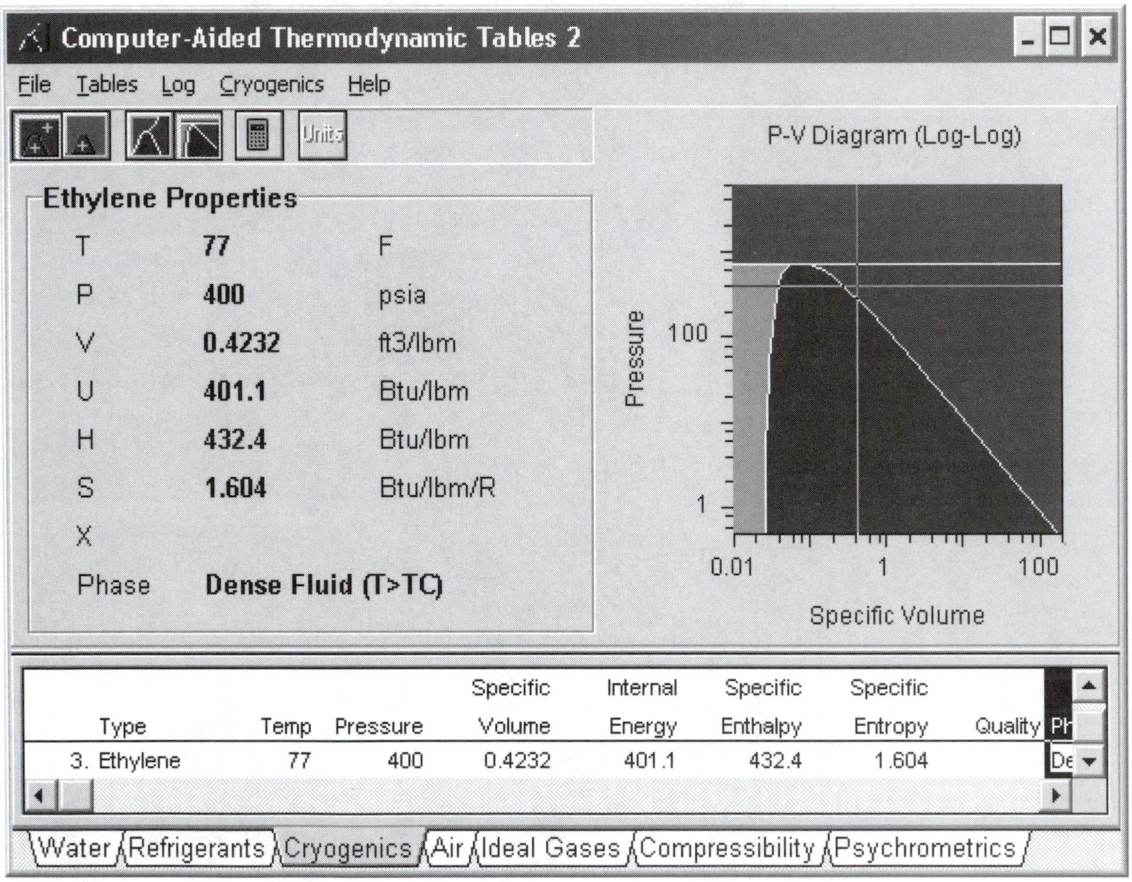

Here we see that

$$v = 0.4232\ \text{ft}^3/\text{lbm}$$

which is 18% smaller than for ideal gas. The difference is higher for a higher pressure.

CHAPTER 4
STUDY PROBLEMS
WORK AND HEAT

- **Work and its units**
- **Boundary work**
- **Other types of work**
- **Heat transfer**

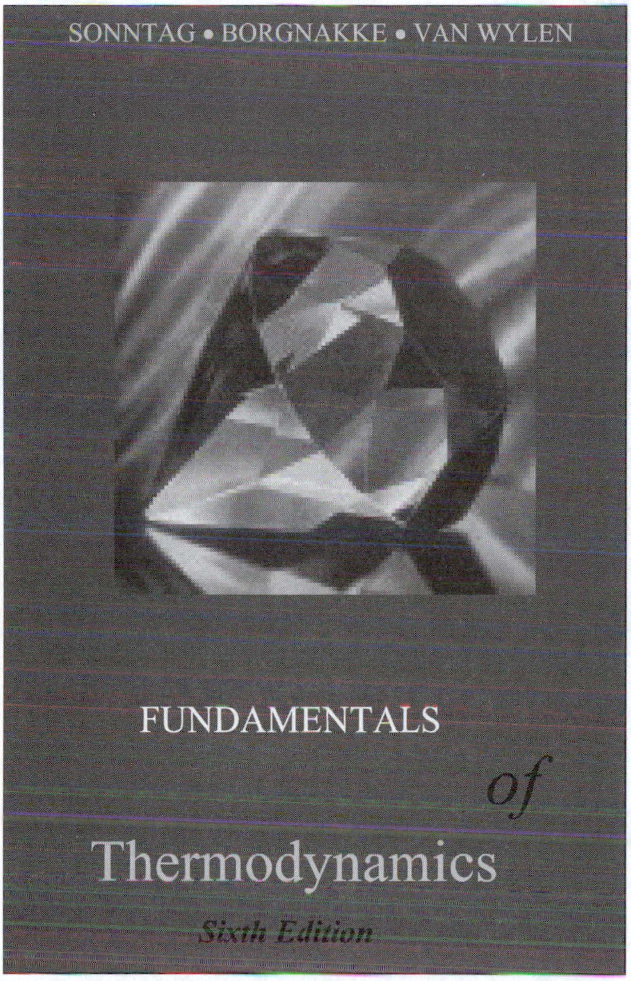

4.1 Energy absorbed in a brake pad

A brake system generates a friction force of 2400 N between a moving and a stationary surface. Assume the surface is moving at 75 m/s and a sliding distance of the braking surface of 35 m to a complete standstill. How much work did the brake system dissipate, and what is the power absorbed at the beginning of the process?

Solution:

We assume the friction force is constant and acting over a distance of 35 m. This gives a work of

$$W = \int F\,dx = F\,L = 2400\text{ N} \times 35\text{ m} = \mathbf{84\,000\ J}$$

The instantaneous power is

$$\text{Power} = \dot{W} = F\,\mathbf{V} = 2400\text{ N} \times 75\text{ m/s} = \mathbf{180\,000\ W = 180\ kW}$$

With the constant force the work is evenly spread out over the 35 m distance, but not in time. As the velocity decreases power is lower and eventually drops to zero.

Brake shoes for a drum brake. The drum is fixed to the rotating wheel and the shoes are stationary and are pressed out by a hydraulic piston.

4.2 Work in a constant pressure process

A piston/cylinder setup contains 0.5 kg of water at 500 kPa, 200°C, which is then cooled at constant pressure so all the water ends up at ambient temperature of 20°C. Find the total work done by the water and show the process in a P-v and a T-v diagram.

Solution:

A constant pressure gives a constant force on the piston so the work is

$$ {}_1W_2 = \int_1^2 F\,dx = \int_1^2 P\,dV = P\,(V_2 - V_1) $$

We must find the beginning and end state properties to get the volumes.

State 1: At 500 kPa in Table B.1.2 Tsat = 151.9°C < T so this state is superheated vapor and we proceed to Table B.1.3
The specific volume: $v_1 = 0.42492\ m^3/kg$ in the 500 kPa section.

State 2: At 500 kPa in Table B.1.2 Tsat = 151.9°C > T so this state is subcooled liquid (compressed liquid) and we proceed to Table B.1.4
The specific volume: $v_2 = 0.001002\ m^3/kg$ in the 500 kPa section.
Notice saturated liquid at 20°C from B.1.1 is a very good approximation

$$ {}_1W_2 = P\,(V_2 - V_1) = P\,m\,(v_2 - v_1) $$
$$ = 500\ kPa \times 0.5\ kg \times (0.001002 - 0.42492)\ m^3/kg $$
$$ = -105.98\ (kN/m^2) \times m^3 = -105.98\ kN\ m $$
$$ = \mathbf{-105.98\ kJ} $$

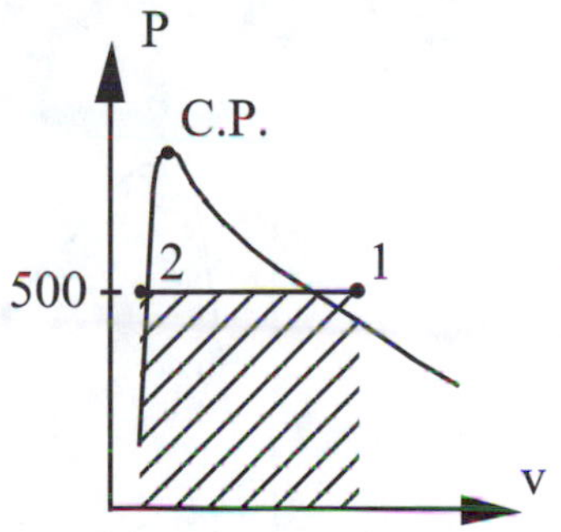

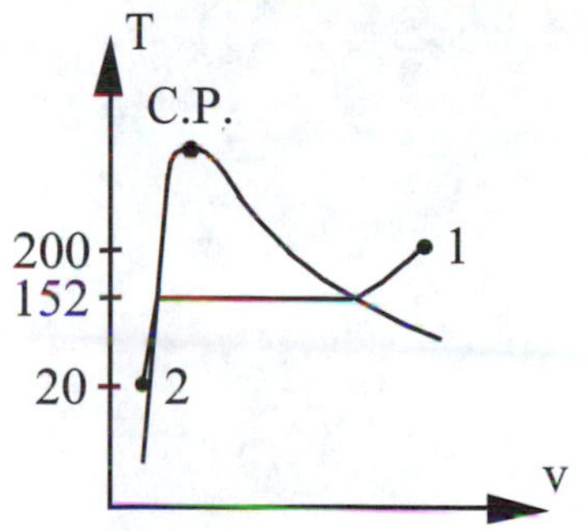

The area in the P-v diagram is the specific work ${}_1w_2 = P\,(v_2 - v_1)$. If we had plotted it in a P-V diagram the area would be ${}_1W_2 = P\,(V_2 - V_1) = m\ {}_1w_2$.

4.3 A linearly decreasing pressure

Consider a piston/cylinder of area 0.025 m^2 containing 0.05 kg of saturated water vapor at 1600 kPa. The piston moves a load in such a manner that the force decreases linearly as the volume goes up $P = P_1 - C(v - v_1)$, $C = 3400$ kPa kg/m^3. The motion ends with a pressure of 400 kPa. Find the maximum force, the displacement (Δx) of the load and the work in the process.

Solution:

The maximum force must be for max P which is at the beginning of the process

$$F_1 = P_1 A = 1600 \text{ kPa} \times 0.025 \text{ m}^2 = \mathbf{40\ kN}$$

Find state 2 from the process equation and the final pressure

$$v_2 = v_1 + (P_1 - P_2) / C = 0.1238 + (1600 - 400)/3400 = 0.47674 \text{ m}^3/\text{kg}$$

From table B.1.3 and state 2 as (400 kPa, 0.47674 m^3/kg) we have approximately T = 155°C.

$$\Delta x = m(v_2 - v_1)/A = 0.05\ (0.47674 - 0.1238)/\ 0.025 = \mathbf{0.706\ m}$$

$$_1W_2 = \int_1^2 P\,dV = m \int_1^2 P\,dv = m \times \text{area} = m\frac{1}{2}(P_1 + P_2)(v_2 - v_1)$$

$$= \frac{1}{2} \times 0.05 \text{ kg } (1600 + 400) \text{ kPa } (0.47674 - 0.1238) \text{ m}^3/\text{kg} = \mathbf{17.65\ kJ}$$

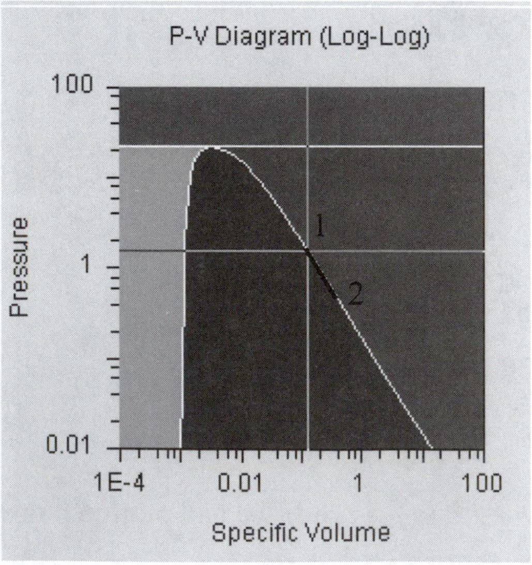

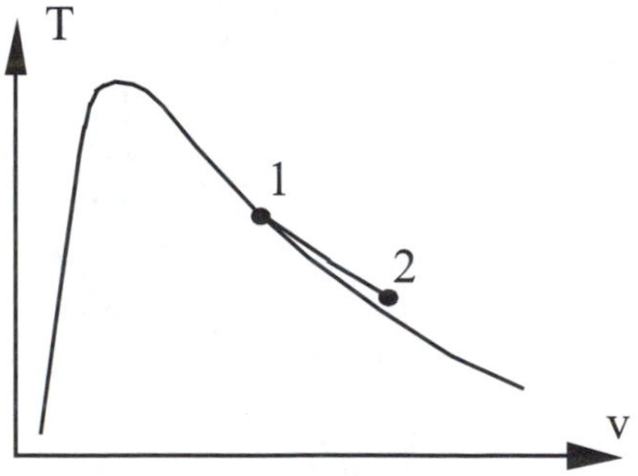

The area below the curve in the P-v diagram is $_1w_2$ if both the axis are linear.

4.4 A polytropic process

Hot hydrogen gas at 800 K expands in a piston/cylinder from 500 kPa to 125 kPa. The process is polytropic with n = 1.5. We want to find the final temperature, the volume expansion ratio v_2/v_1, the specific work in the process and show the process in a P-v diagram.

Solution:

The hydrogen gas is an ideal gas at these states (refer to Table A.2)
State 1: 800 K, 500 kPa
State 2: ? , 125 kPa

Process: $Pv^n = C$
Table A.5: R = 4.1243 kJ/kg K

From the process equation we have $P_2v_2^n = Pv_1^n$ giving

$$v_2 / v_1 = (P_2/P_1)^{1/n} = \left(\frac{500}{125}\right)^{1/1.5} = 4^{2/3} = \mathbf{2.52}$$

then from the equation of state $P_2v_2/T_2 = P_1v_1/T_1$ so we solve for T_2

$$T_2 = T_1 \,(P_2/P_1)\,(v_2/v_1) = 800\,(\frac{125}{500})\,2.52 = \mathbf{504\ K}$$

The work is integrated as in equation 4.4 and then ideal gas law Pv = RT is used

$$_1w_2 = \frac{1}{1\text{-}n}(P_2v_2 - P_1v_1) = \frac{R}{1\text{-}n}(T_2 - T_1)$$

$$= \frac{4.1243}{1\text{-}1.5}(504 - 800) = \mathbf{2442\ kJ/kg}$$

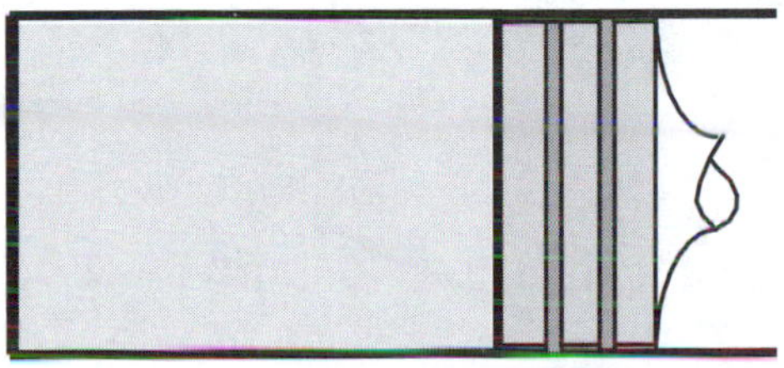

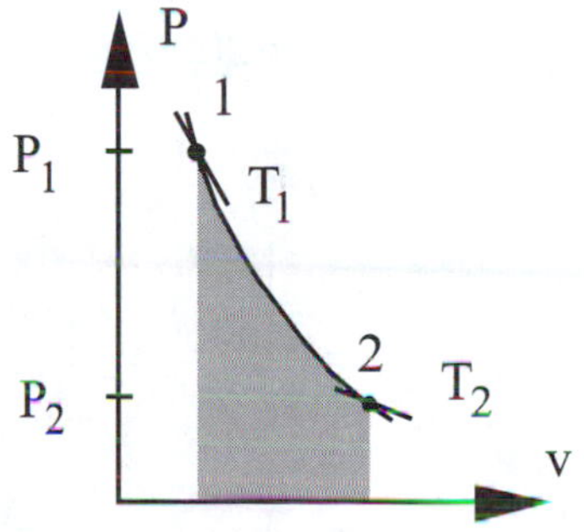

4.5 A possible multistep process

A piston cylinder contains 0.5 kg air at 1000 K and 2000 kPa. There is a constant force on the piston and the minimum volume is 30 L if the piston is at the stops. Now the air cools by heat transfer to a final temperature of 500 K. We want to know the final volume and the work during the process.

We recognize this is a possible two-step process, one of constant P and one of constant V. This behavior is dictated by the construction of the device.

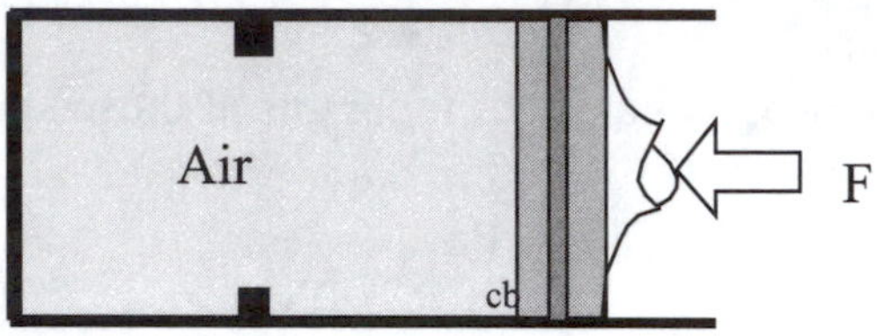

Process: $P = \text{constant} = F/A = P_1 \quad \text{if} \quad V > V_{min}$

$V = \text{constant} = V_{1a} = V_{min} \quad \text{if} \quad P < P_1$

State 1: (P, T) $V_1 = mRT_1/P_1 = 0.5 \times 0.287 \times 1000/2000 = 0.07175 \text{ m}^3$

The only possible P-V combinations for this system is shown in the diagram so both state 1 and 2 must be on the two lines. For state 2 we need to know if it is on the horizontal P line segment or the vertical V segment. Let us check state 1a:

State 1a: $P_{1a} = P_1,\ V_{1a} = V_{min}$

$$\text{Ideal gas so } T_{1a} = T_1 \frac{V_{1a}}{V_1} = 1000 \times \frac{0.03}{0.07175} = 418 \text{ K}$$

We see that $T_2 > T_{1a}$ and state 2 must have $P_2 = P_{1a} = P_1 = 2000$ kPa.

$$V_2 = V_1 \times \frac{T_2}{T_1} \times \frac{P_1}{P_2} = 0.07175 \times \frac{500}{1000} \times \frac{2000}{2000} = 0.035875 \text{ m}^3$$

The work is the area under the process curve in the P-V diagram

$${}_1W_2 = \int_1^2 P\, dV = P_1 (V_2 - V_1)$$

$$= 2000 \text{ kPa } (0.035875 - 0.07175) \text{ m}^3 = \mathbf{-\ 71.75\ kJ}$$

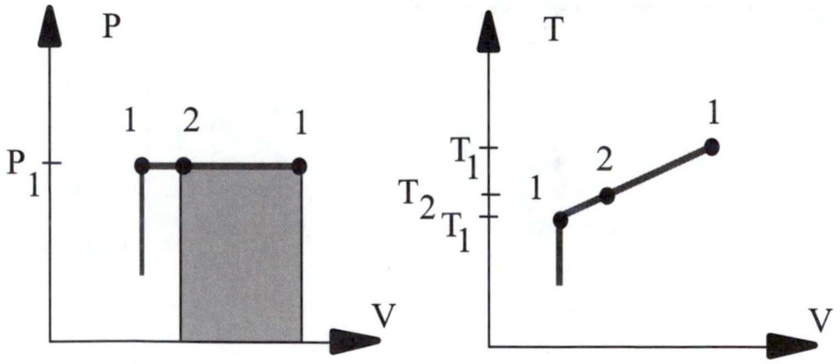

4.6 A disc brake system

Two brake pads in a disc brake system generate a normal force of $F_n = 1$ kN. Assume a friction coefficient of $\mu = 0.3$ so the friction force is $F = \mu\ F_n$ and that the disc speed at that location is 15 m/s. How much power does that brake system dissipate?

Solution:

Work is force times displacement	$dW = F\ dx$
Power is force times rate of displament	$\dot{W} = F\ \mathbf{V}$

The force is from the normal force and the friction (dynamic) coefficient

$$F = \mu\ F_n = 0.3 \times 1\,000\ \text{N} = 300\ \text{N}$$

$$\dot{W} = F\ \mathbf{V} = 300\ \text{N} \times 15\ \text{m/s} = 4\,500\ \text{J/s} = \mathbf{4.5\ kW}$$

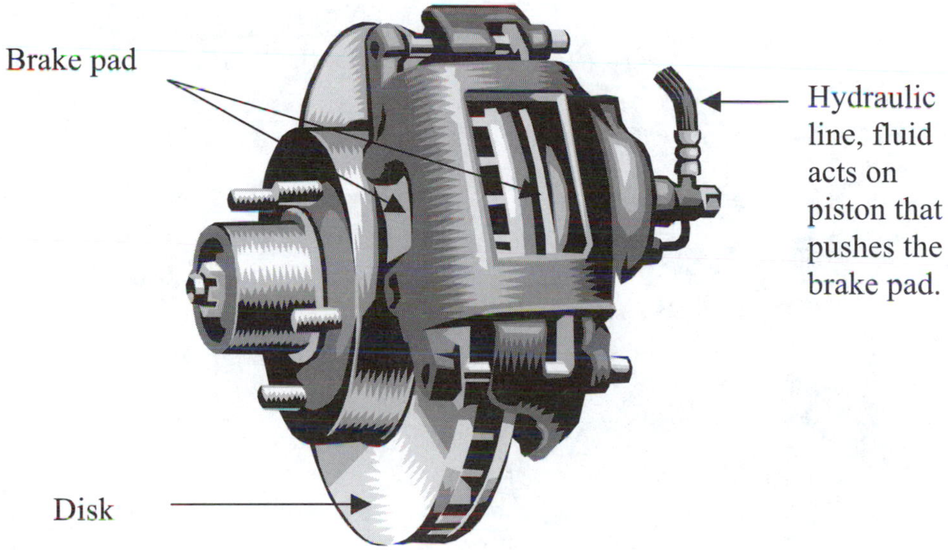

4.7 Conduction and convection in an electric stove

A rate of energy transfer of 0.5 kW from an electric heating element goes through a 0.5 cm thick plate of conductivity k = 1 W/m K and then passes through a convection layer with h = 80 W/m^{2}K. Assume an area of 0.025 m^2 and find the temperature difference across each of the two layers and the energy transferred over a period of 10 minutes.

Solution:

This is a steady state conduction through the plate for which we have

$$\dot{Q} = - kA \frac{dT}{dx} = kA \frac{\Delta T}{\Delta x}$$

$$\Delta T_{conduction} = \frac{\dot{Q}\,\Delta x}{kA} = \frac{500 \text{ W} \times 0.005 \text{ m}}{1 \text{ (W/m K)} \times 0.025 \text{ m}^2} = \mathbf{100\ K}$$

The steady state convection layer follows Newton's law of cooling, $\dot{Q} = hA\,\Delta T$, so the temperature difference is

$$\Delta T_{convection} = \frac{\dot{Q}}{hA} = \frac{500 \text{ W}}{80 \text{ (W/m}^2\text{K)} \times 0.025 \text{ m}^2} = \mathbf{250\ K}$$

The total energy over a time period is, assuming a constant rate,

$$Q = \int \dot{W}_{el.}\, dt = \dot{W}_{el.}\, \Delta t$$

$$= 500 \text{ W} \times 10 \text{ min} \times 60 \text{ s/min} = 300\,000 \text{ J} = \mathbf{300\ kJ}$$

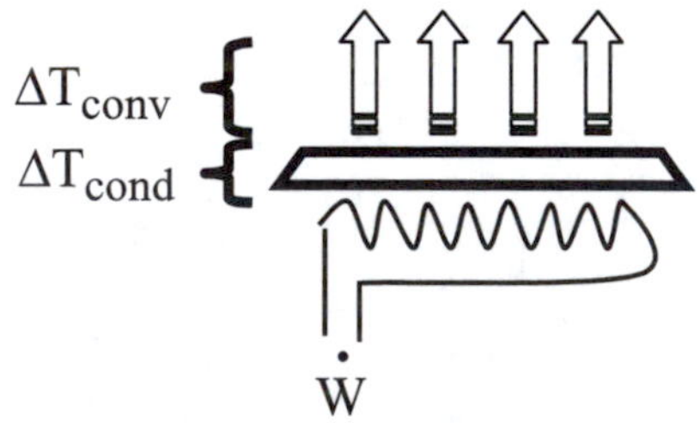

4.8 A radiation and convection example

In a furnace we have two walls of area 1 m^2 facing each other with a flow of air at 400 K between them. Both surfaces have emissivity of 0.8 and the left surface is at 1000 K while the right surface is at T_s and the flow has a convection heat transfer coefficient of h = 50 W/m^2K. We want to find the net radiative heat transfer rate and the convective heat transfer rate as a function of the surface temperature T_s for the range 400-1000 K.

Solution:

The left surface has also convection to the air, but that does not affect this problem as air T is given.

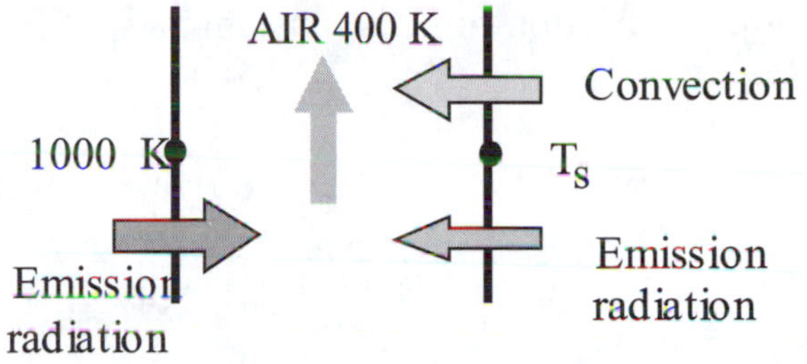

The rate of heat transfer by convection is from Newtons law of cooling

$$\dot{Q}_{convection} = hA\,\Delta T = hA\,(T_s - T_{fluid})$$

with a sign so it leaves the surface and enters the fluid. The rate of net radiation heat transfer from the hot side to the surface is

$$\dot{Q}_{radiation} = \varepsilon\sigma A\,(\,T_{left}^4 - T_s^4\,)\,.$$

The two rates are computed as a function of T_s as shown in the following table also plotted in the figure below. The convection varies linearly with the surface temperature whereas the radiation being non-linear drops readily towards zero as the surface temperature approaches the left side temperature of 1000 K.

T_s [K]	400	425	450	500	600	700	800	900
$\dot{Q}_{convection}$ [kW]	0	1.25	2.5	5	10	15	20	25
$\dot{Q}_{radiation}$ [kW]	44.2	43.9	43.5	42.5	39.5	34.5	26.8	15.6

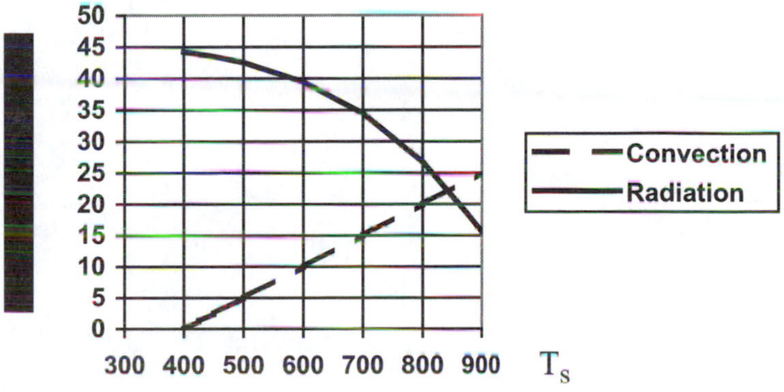

4.1E Energy absorbed in a brake pad

A brake system generates a friction force of 600 lbf between a moving and a stationary surface. Assume the surface is moving at 250 ft/s and a sliding distance of the braking surface of 120 ft to a complete standstill. How much work did the brake system dissipate, and what is the power absorbed at the beginning of the process?

Solution:

We assume the friction force is constant and acting over a distance of 35 m. This gives a work of

$$W = \int F\,dx = F\,L = 600\text{ lbf} \times 120\text{ ft} = \mathbf{72\,000\text{ lbf-ft} = 92.5\text{ Btu}}$$

From Table A.1 we have 1 Btu = 778.169 lbf-ft.
The instantaneous power is

$$\text{Power} = \dot{W} = F\,\mathbf{V} = 600\text{ lbf} \times 250\text{ ft/s} = \mathbf{150\,000\text{ lbf-ft/s} = 193\text{ Btu/s}}$$

With the constant force the work is evenly spread out over the 120 ft distance, but not in time. As the velocity decreases power is lower and eventually drops to zero.

Brake shoes for a drum brake. The drum is fixed to the rotating wheel and the shoes are stationary and are pressed out by a hydraulic piston.

4.2E Work in a constant pressure process

A piston/cylinder setup contains 1 lbm of water at 80 psia, 400 F, which is then cooled at constant pressure so all the water ends up at ambient temperature of 80 F. Find the total work done by the water and show the process in a P-v and a T-v diagram.

Solution:

A constant pressure gives a constant force on the piston so the work is

$$ {}_1W_2 = \int_1^2 F\,dx = \int_1^2 P\,dV = P\,(V_2 - V_1) $$

We must find the beginning and end state properties to get the volumes.

State 1: At 80 psia in Table F.7.1 Tsat = 312.06 F < T so this state is superheated vapor and we proceed to Table F.7.2
The specific volume: $v_1 = 6.217\ \text{ft}^3/\text{lbm}$ in the 80 psia section.

State 2: At 80 psia in Table F.7.1 Tsat = 312.06 F > T so this state is subcooled liquid (compressed liquid) and we proceed to Table F.7.1 as the pressure is too low to be found in the compressed liquid Table F.7.3. We use saturated liquid same T.
The specific volume: $v_2 = 0.01607\ \text{ft}^3/\text{lbm}$ at 80 F.

$$ {}_1W_2 = P\,(V_2 - V_1) = P\,m\,(v_2 - v_1) $$
$$ = 80\ \text{lbf} \times 1\ \text{lbm} \times (0.01607 - 6.217)\ \text{ft}^3/\text{lbm} $$
$$ = \mathbf{-\,71\,435\ lbf\text{-}ft} $$
$$ = \mathbf{-\,91.8\ Btu} $$

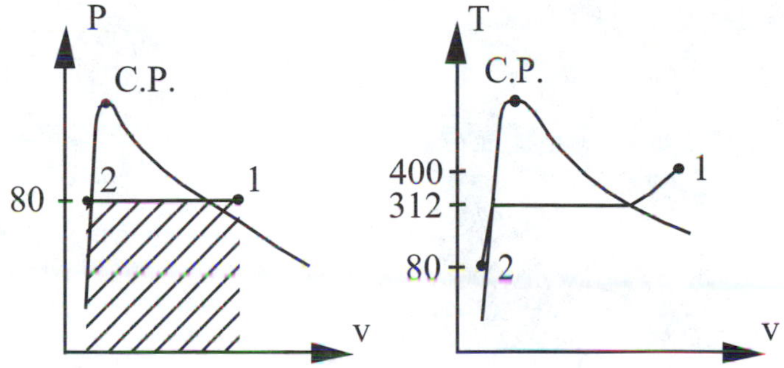

The area in the P-v diagram is the specific work ${}_1w_2 = P\,(v_2 - v_1)$. If we had plotted it in a P-V diagram the area would be ${}_1W_2 = P\,(V_2 - V_1) = m\,{}_1w_2$.

4.7E Conduction and convection in an electric stove

A rate of energy transfer of 0.5 kW from an electric heating element goes through a 0.2 in thick plate of conductivity k = 0.5 Btu/h-ft-R and then passes through a convection layer with h = 14 Btu/h-ft^2-R. Assume an area of 0.3 ft^2 and find the temperature difference across each of the two layers and the energy transferred over a period of 10 minutes.

Solution:

This is a steady state conduction through the plate for which we have

$$\dot{Q} = - kA \frac{dT}{dx} = kA \frac{\Delta T}{\Delta x}$$

The heat transfer becomes with conversion from Table A.1

$$\dot{Q} = \dot{W}_{el.} = 500 \text{ W} = 500 \text{ W} \times 3.412 \frac{\text{Btu}}{\text{W-h}} = 1706 \text{ Btu/h}$$

$$\Delta T_{conduction} = \frac{\dot{Q}\,\Delta x}{kA} = \frac{1706 \text{ Btu/h} \times 0.2 \text{ in}}{0.5 \text{ (Btu/h-ft-R)} \times 0.3 \text{ ft}^2 \times 12 \text{ (in/ft)}} = \mathbf{190\ R}$$

The steady state convection layer follows Newton's law of cooling, $\dot{Q} = hA\,\Delta T$, so the temperature difference is

$$\Delta T_{convection} = \frac{\dot{Q}}{hA} = \frac{1706 \text{ Btu/h}}{14 \text{ (Btu/h-ft}^2\text{-R)} \times 0.3 \text{ ft}^2} = \mathbf{406\ R}$$

The total energy over a time period is, assuming a constant rate,

$$Q = \int \dot{W}_{el.}\, dt = \dot{W}_{el.}\, \Delta t = 1706 \frac{\text{Btu}}{\text{h}} \times \frac{10}{60} \text{h} = \mathbf{284\ Btu}$$

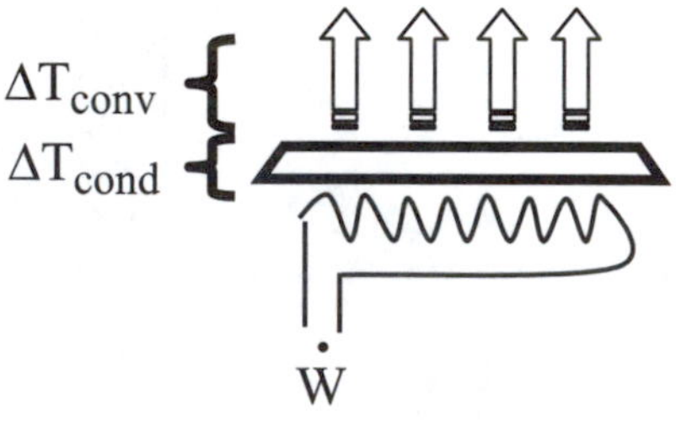

CHAPTER 5
STUDY PROBLEMS

THE FIRST LAW OF THERMODYNAMICS
THE ENERGY EQUATION

- **The energy equation**
- **Internal energy, enthalpy and specific heats**
- **Energy equation: solids and liquids**
- **Energy equation: Ideal gas**
- **Energy equation in rate form**

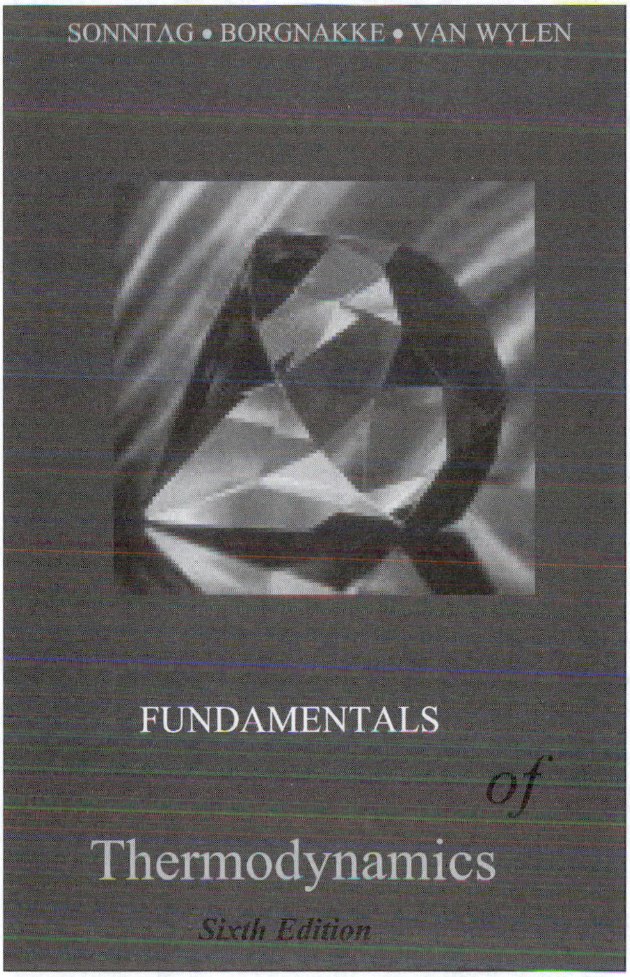

5.1 The work term and stored energy

Consider the setup in Example 4.2 where we found the work term in an expansion process. The energy that is going out is ${}_1W_2$, how much is stored where?

Solution:

Consider the piston as a control volume, which is a control mass with negligible velocity.

Energy Eq.: $E_2 - E_1 = U_2 - U_1 + m_p g(x_2 - x_1) = {}_1W_2 - W_{F1} - W_{spring} - W_{atm}$

The piston receives work from the mass inside the cylinder and gives out work to the force F_1, to the spring and to the atmosphere. The balance is stored in the piston in this case as potential energy assuming the piston does not change U.

We can find the other work terms

Single point force: $W_{F1} = \int F_1\, dx = F_1\,(x_2 - x_1) = F_1\,(V_2 - V_1)/A$

This energy is stored in the sub-system that provides the force F_1.

The spring: $W_{spring} = \int F_{sp}\, dx = \int k_s(x - x_o)\, dx$

$$= \frac{1}{2} k_s\,[(x_2 - x_o)^2 - (x_1 - x_o)^2]$$

This is an increase in the spring internal energy (you may also call this potential energy).

The atmosphere: $W_{atm} = \int P\, dV = P_o\,(V_2 - V_1)$

This is the increase in the atmosphere internal energy (it is being compressed).

We therefore realize the work can be written from the energy equation as:

$$\begin{aligned} {}_1W_2 &= (E_2 - E_1)_p + W_{F1} + W_{spring} + W_{atm} \\ &= (E_2 - E_1)_p + (E_2 - E_1)_{F1} + (E_2 - E_1)_{spring} + (E_2 - E_1)_{atm} \end{aligned}$$

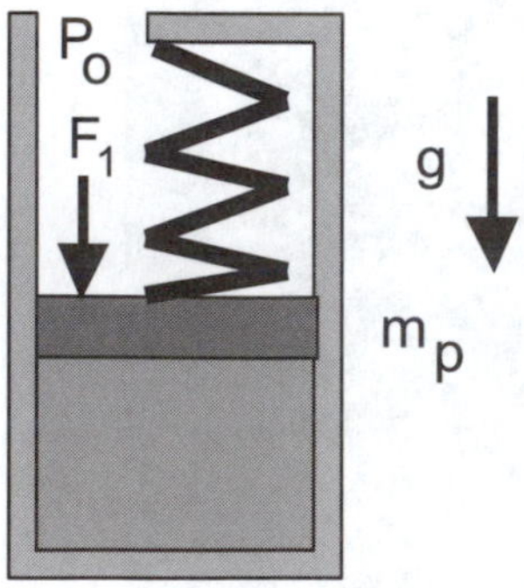

5.2 Table look-up involving u and h

Find the missing properties among (*P, T, v, u, h*) together with *x* if applicable and give the phase of the substance.

a.	H_2O,	$P = 1000$ kPa, $v = 0.02$ m^3/kg
b.	H_2O,	$T = -8°C$, $P = 100$ kPa
c.	R-134a,	$T = 40°C$, $u = 300$ kJ/kg
d.	R-12,	$T = 10°C$, $u = 177.5$ kJ/kg

Solution:

a) Table B.1.1 at 1000 kPa: $v_f < v < v_g \Rightarrow$ **L+V**, $T = T_{sat} =$ **179.91°C**

$$x = \frac{v - v_f}{v_{fg}} = \frac{0.02 - 0.001127}{0.19332} = \mathbf{0.09762}$$

$$u = u_f + x\, u_{fg} = 761.67 + 0.09762 \times 1821.97 = \mathbf{939.53\ kJ/kg}$$

$$h = h_f + x\, h_{fg} = u + Pv = 939.53 + 1000 \times 0.02 = \mathbf{959.53\ kJ/kg}$$

b) Table B.1.1 : $T < T_{triple\ point}$ => B.1.5: $P > P_{sat}$ so **compressed solid**

$v \cong v_i =$ **1.089×10^{-3} m^3/kg** ; $u \cong u_i =$ **-350.02 kJ/kg** $h \cong h_i =$ **-350.02 kJ/kg**

approximate compressed solid with saturated solid properties at same T.

c) Table B.5.1: $u < u_g = 399.46$ kJ/kg => **L+V mixture**, P = **1017 kPa**

$$x = \frac{u - u_f}{u_{fg}} = \frac{300 - 255.65}{143.81} = \mathbf{0.3084},$$

$$v = v_f + x\, v_{fg} = 0.000873 + 0.3084 \times 0.01915 = \mathbf{0.00678\ m^3/kg},$$

$$h = h_f + x\, h_{fg} = 256.54 + 0.3084 \times 163.28 = \mathbf{306.9\ kJ/kg}$$

(we could also have done h = u + Pv)

d) Table B.3.1: $u > u_g$ => **superheated vapor,** look in B.3.2 at 10°C very close at:

P = **200 kPa**, v = **0.09255 m^3/kg,** h = **196.02 kJ/kg**

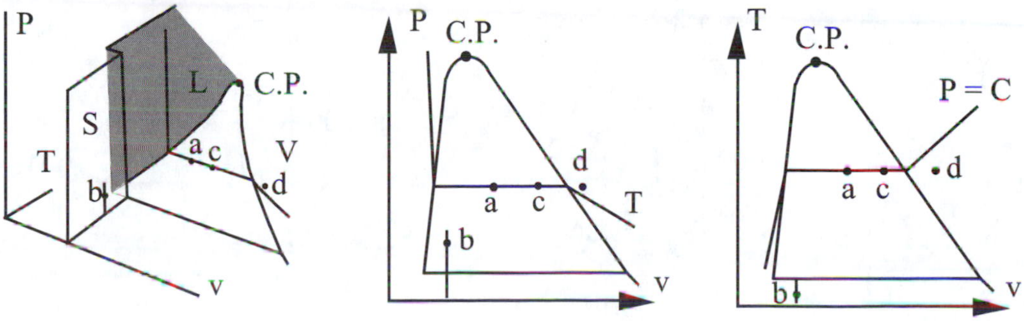

5.3 A two step process for a control mass

A piston/cylinder device contains 0.2 kg water at 1000 kPa, 300°C. There is a constant force on the piston with stops mounted so a minimum volume is 0.025 m^3. It is now cooled to 100°C. Find the final water volume, the work and heat transfer in the overall process and show the process in P-v and T-v diagrams.

Solution:

C.V. Water, this is a control mass. Process determined by device behavior.

Energy Eq.: $U_2 - U_1 = m(u_2 - u_1) = {}_1Q_2 - {}_1W_2$

State 1: (P, T) Table B.1.3: $v_1 = 0.25794\ m^3/kg$, $u_1 = 2793.21$ kJ/kg

State 2: (T, ?)

Process: $P = \text{constant} = F/A = P_1 \quad \text{if} \quad V > V_{min}$

$V = \text{constant} = V_{1a} = V_{min} \quad \text{if} \quad P < P_1$

The only possible P-v combination for this system is shown in the diagram so both state 1 and 2 must be on the two lines. For state 2 we need to know if it is on the horizontal P line segment or the vertical v segment. Let us check state 1a:

State 1a: $P,\ v_{1a} = V_{min}/m = \frac{0.025}{0.2} = 0.125\ m^3/kg$ from Table B.1.2

$v_f < v_{1a} < v_g$ so two-phase and $T_{1a} = T_{sat} = 179.9°C$

We see that $T_2 < T_{1a}$ and state 2 must have $v_2 = v_{1a} = v_{min} = 0.125\ m^3/kg$.

$$x_2 = \frac{v_2 - v_f}{v_{fg}} = \frac{0.125 - 0.001044}{1.67185} = 0.07414$$

$$u_2 = u_f + x_2\, u_{fg} = 418.91 + x_2\, 2087.58 = 573.69 \text{ kJ/kg}$$

The work is the area under the process curve in the P-V diagram

$$_1W_2 = m\ _1w_2 = \int_1^2 P\, dV = m \int_1^2 P\, dv = mP_1\,(v_{1a} - v_1) + 0$$

$$= 0.2 \text{ kg} \times 1000 \text{ kPa } (0.125 - 0.25794)\ m^3/kg = \mathbf{-\ 26.6\ kJ}$$

Now the heat transfer is found from the energy equation

$$_1Q_2 = m(u_2 - u_1) + {}_1W_2 = 0.2\,(573.69 - 2793.21) - 26.6 = \mathbf{-470.5\ kJ}$$

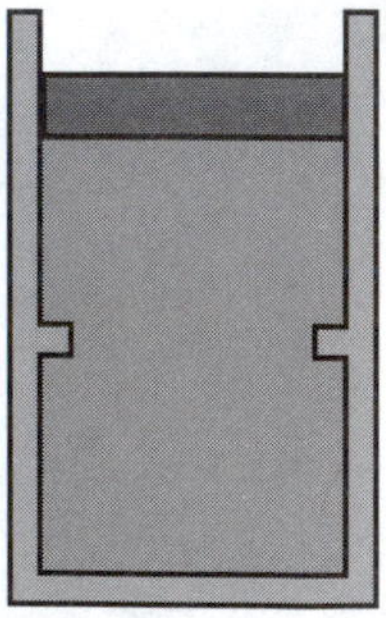

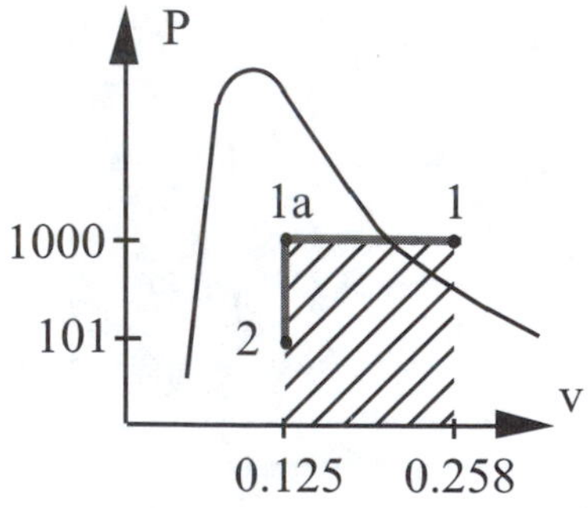

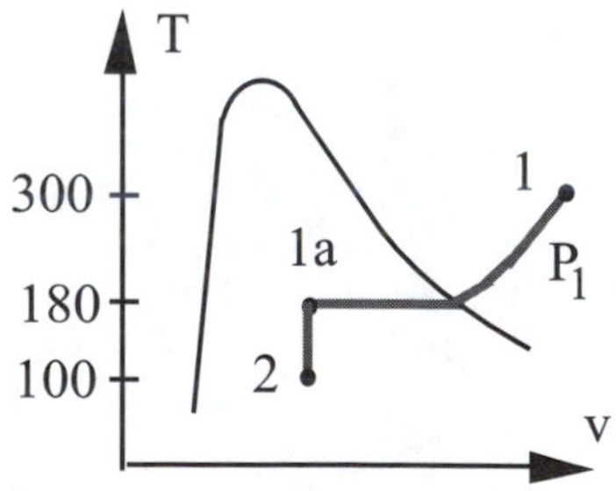

5.4 Mixing of R-134a at two different states

A piston cylinder with a constant force on it contains 0.25 kg R-134a saturated vapor in volume A and 0.1 kg of R-134a at 60°C in volume B both at 600 kPa. The two masses mix in an adiabatic process. We want to know the final temperature and the work in the process.

Solution:

C.V. All the R-134a in volume A and B, adiabatic so Q = 0.

Continuity Eq.: $m_2 - m_A - m_B = 0$

Energy Eq.5.11: $m_2u_2 - m_Au_A - m_Bu_B = -{}_1W_2$

Process: P = Constant => ${}_1W_2 = \int PdV = P(V_2 - V_1)$

Substitute the work term into the energy equation and rearrange to get

$$m_2u_2 + P_2V_2 = m_2h_2 = m_Au_A + m_Bu_B + PV_1 = m_Ah_A + m_Bh_B$$

where the last rewrite used $PV_1 = PV_A + PV_B$.

State A1: Table B.5.2 $h_A = 410.66$ kJ/kg ; $v_A = 0.03442$ m^3/kg

State B1: Table B.5.2 $h_B = 448.28$ kJ/kg ; $v_B = 0.04145$ m^3/kg

Energy equation gives:

$$h_2 = \frac{m_A}{m_2} h_A + \frac{m_B}{m_2} h_B = \frac{0.25}{0.35} 410.66 + \frac{0.1}{0.35} 448.28 = 421.41 \text{ kJ/kg}$$

State 2: (P_2, h_2) ⇒ T_2 = **32.4°C,** $v_2 = 0.03653$ m^3/kg

$${}_1W_2 = \int PdV = P(V_2 - V_1)$$

$$= 600 \text{ kPa } (0.35 \times 0.03653 - 0.25 \times 0.03442 - 0.1 \times 0.04145) \text{ m}^3$$

$$= \mathbf{0.0213 \text{ kJ}}$$

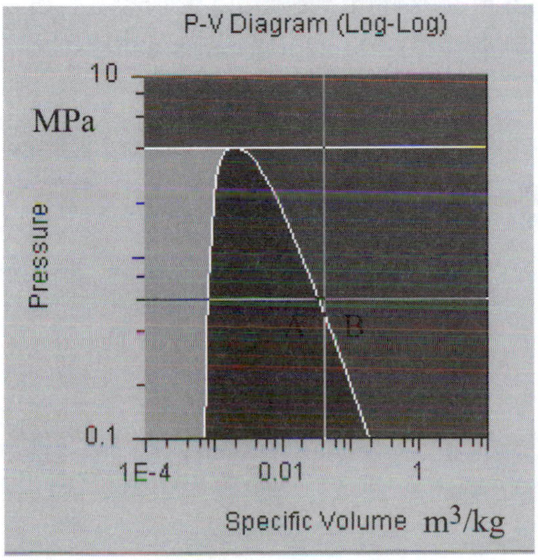

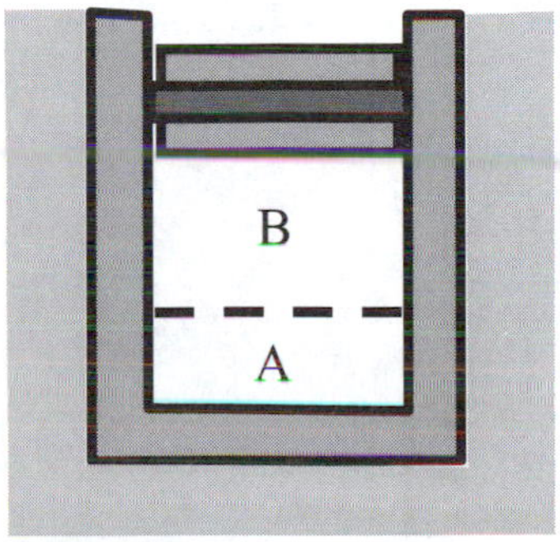

5.5 An isothermal expansion process

A mass of 1 kg of air contained in a cylinder at 1.5 MPa, 1000 K, expands in a reversible isothermal process to a pressure 5 times smaller. Calculate the heat transfer and the work during the process.

Solution:

C.V. Air, which is a control mass.

Energy Eq.5.11: $m(u_2 - u_1) = {}_1Q_2 - {}_1W_2$

Process: $T = \text{constant}$

For an ideal gas constant T => $u_2 = u_1$ then ${}_1Q_2 = {}_1W_2$

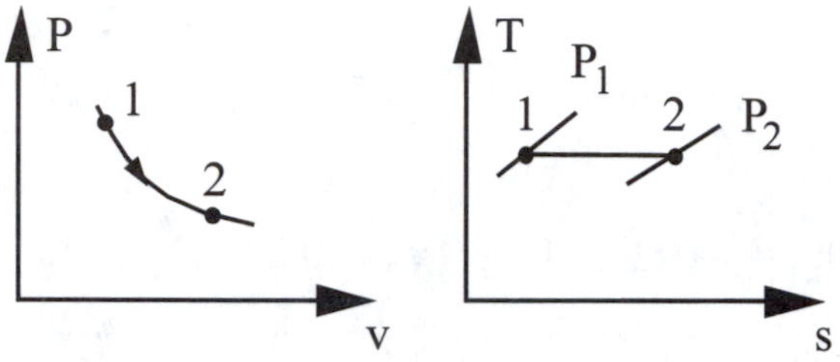

From the process equation and ideal gas law

$$PV = mRT = \text{constant}$$

we can calculate the work term as in Eq.4.5

$${}_1Q_2 = {}_1W_2 = \int PdV = P_1V_1 \ln (V_2/V_1) = mRT_1 \ln (V_2/V_1)$$

$$= 1 \times 0.287 \times 1000 \ln (5) = \mathbf{461.91\ kJ}$$

Comment: Notice this is a polytropic process since the substance is an ideal gas. If it had not been an ideal gas it would not be a polytropic process and we could not have found the work this way. The solution is then done with entropy in chapter 8, see study 8-9.

5.6 A polytropic process

A piston/cylinder setup contains 0.5 kg of air at 20°C and 100 kPa. The pressure is now increased to a final pressure of 600 kPa during which the air goes through a polytropic process with exponent n = 1.3.

a) Find the initial volume. b) Plot the P-v diagram for the process
c) Find the final temperature. d) Find the work in the process

Solution:

Continuity eq.: $m_2 = m_1 = m$;

Energy eq.: $m(u_2 - u_1) = {}_1Q_2 - {}_1W_2$

Process eq.: $Pv^n = C$ or $PV^n = C$

State 1: Air is an ideal gas, $Pv = RT$

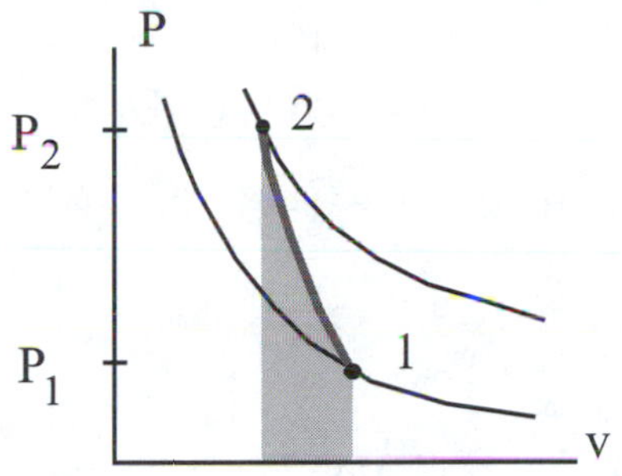

$$v_1 = RT_1 / P_1 = 0.287 \times 293.15 / 100 = 0.84134 \ \ m^3/kg$$

$$V_1 = mv_1 = 0.5 \times 0.84134 = \mathbf{0.4207\ m^3}$$

State 2: $P_2 = 600$ kPa and on process line =>

$$P_2v_2^n = P_1v_1^n \quad => \quad v_2 = v_1 \left(\frac{P_1}{P_2}\right)^{1/n} = v_1 \left(\frac{100}{600}\right)^{0.76923}$$

$$v_2 = 0.84134 \times 0.252 = 0.212 \ m^3/kg$$

$$T_2 = \frac{P_2v_2}{R} = \frac{600 \text{ kPa} \times 0.212 \text{ m}^3}{0.287 \text{ kJ/kg K}} = \mathbf{443.2\ K}$$

If you manipulate the formula for the process and substitute the ideal gas law you could also have expressed T_2 directly as

$$T_2 = T_1 \left(\frac{P_2}{P_1}\right)^{1 - 1/n} = T_1 \left(\frac{100}{600}\right)^{0.23076} = 293.15 \times 1.512 = 443.2 \text{ K}$$

The work term is from Eq.(4.4)

$${}_1W_2 = \frac{1}{1 - n} m\,(P_2v_2 - P_1v_1) = \frac{1}{1 - 1.3} 0.5\,(\,600 \times 0.212 - 100 \times 0.84134)$$

$$= \mathbf{-\ 71.8\ kJ}$$

5.7 Heat exchange between solid and gas

A silicon wafer of volume 0.1 L is heat treated at 250°C and now cooled in a 200-L closed air box initially at 20°C, 101 kPa. The box itself consists of 2 kg stainless 304 steel which has the same temperature as the air. Assuming no heat transfer with the surroundings, what is the final uniform temperature of all the masses? What is the final air pressure?

Solution:

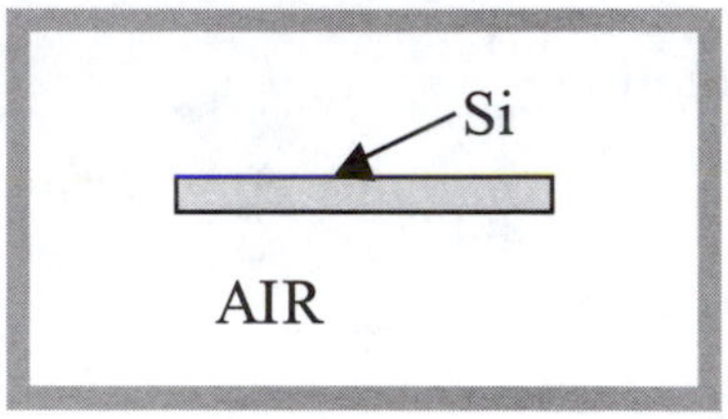

C.V. The rigid steel box, the air and the silicon wafer so this is a control mass.

Process Eq: $V = \text{constant}\ ;\ {}_1W_2 = 0$ and ${}_1Q_2 = 0$

Energy Eq.: $U_2 - U_1 = {}_1Q_2 - {}_1W_2 = 0 - 0 = 0$

$$U_2 - U_1 = m_{steel}(u_2 - u_1)_{steel} + m_{air}(u_2 - u_1)_{air} + m_{si}(u_2 - u_1)_{si}$$

Properties from Tables A.3:

$$C_{steel} = 0.46 \text{ kJ/kg K},\ \ C_{si} = 0.7 \text{ kJ/kg K},\ \ \rho_{si} = 2330 \text{ kg/m}^3$$

$$m_{si} = V\rho = 0.0001 \times 2330 = 0.233 \text{ kg}$$

Table A.5: $C_{V\ air} = 0.717$ kJ/kg K, $R = 0.287$ kJ/kg K

Ideal gas: $m_{air} = \dfrac{PV}{RT} = \dfrac{101 \text{ kPa } (200 - 0.1)\ 10^{-3} \text{ m}^3}{0.287 \times 293.15 \text{ kJ/kg}} = 0.240$ kg

For solids and liquids Eq.5.17, $\Delta u \cong C\ \Delta T$, similar to a gas Eq.5.20: $\Delta u \cong C_V\ \Delta T$,

The energy equation for the C.V. becomes

$$m_{steel}C_{steel}(T_2 - T_{1,steel}) + m_{si}C_{si}(T_2 - T_{1,si}) + m_{air}C_{V\ air}(T_2 - T_{1,air}) = 0$$

Substitute values

$$2 \times 0.46\ (T_2 - 20) + 0.233 \times 0.7\ (T_2 - 250) + 0.24 \times 0.717\ (T_2 - 20) = 0$$

$$1.255\ T_2 - 18.4 - 40.775 - 3.4416 = 0$$

Solve for T_2: $\Rightarrow T_2 = \mathbf{49.9\ °C}$

From the process and the energy equations we have state 2: $(T_2, v_2 = v_1)$. Using then the ideal gas law, $PV = mRT$, for states 1 and 2 gives

$$P_2 = P_1 \frac{V_1}{V_2} \frac{mRT_2}{mRT_1} = 101 \frac{273.15 + 49.9}{273.15 + 20} = \mathbf{111.3\ kPa}$$

5.8 A transient problem with rates of energy

Assume we have a 1 kg aluminum pot with 1 kg liquid water both at 15°C, 100 kPa. The pot is placed on a stove and heated with an input of 500 W. Neglect heat losses and find the time it will take to bring the water to the boiling point.

Solution:

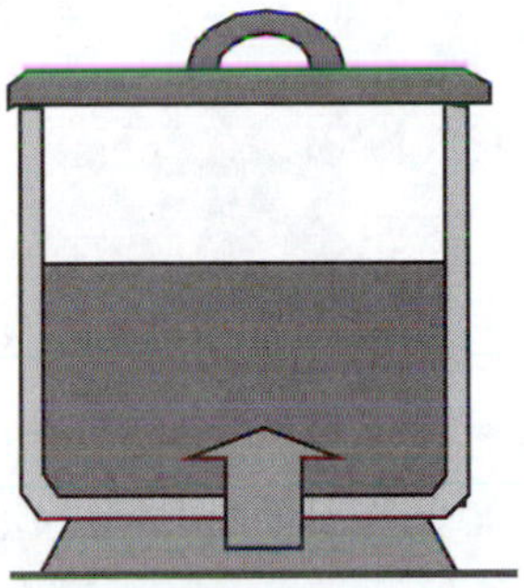

C.V. The pot and water. This is a control mass.

Energy Eq. rate form: $\dot{E} = \dot{Q} - \dot{W} = \dot{Q}$

Energy Eq. integrated: $E_2 - E_1 = {}_1Q_2 = \dot{Q}\,(t_2 - t_1)$

We have no changes in kinetic or potential energy, no change in mass. So

$$E_2 - E_1 = U_2 - U_1 = m_{al}(u_2 - u_1)_{al} + m_{water}(u_2 - u_1)_{water}$$

$$= m_{al}C_{V\ al}(T_2 - T_1) + m_{water}(u_2 - u_1)_{water}$$

$$= 1 \text{ kg} \times 0.9 \text{ kJ/kg K} \times (100 - 15) \text{ K} + 1 \text{ kg} \times (418.91 - 62.98) \text{ kJ/kg}$$

$$= 76.5 + 355.93 = 432.43 \text{ kJ}$$

where we used $C_{V\ al}$ from Table A.3 and Table B.1.1 for $u_{water} = u_f$. Now the time is found as

$$(t_2 - t_1) = {}_1Q_2 / \dot{Q} = \frac{432.43 \times 1000}{500} \text{ s} = 864.86 \text{ s} = \mathbf{14\ min\ 25\ s}$$

Comment: We could also have used $C_{V\ water} = 4.18$ kJ/kg K from Table A.4 that gives

$$(u_2 - u_1)_{water} = C_{V\ water}\,\Delta T = 4.18 \times 85 = 355.3 \text{ kJ/kg}$$

very close to the value found above.

5.2E Table look-up involving u and h

Find the missing properties among (*P, T, v, u, h*) together with *x* if applicable and give the phase of the substance.

a. H_2O, $P = 103$ psia, $v = 1.25$ ft^3/lbm
b. H_2O, $T = 20$ F, $P = 14.7$ psia
c. R-134a, $T = 100$ F, $u = 140$ Btu/lbm
d. R-12, $P = 200$ psia, $h = 131$ Btu/lbm

Solution:

a) Table F.7.1 at 103 psia: $v_f < v < v_g \Rightarrow$ **L+V**, $T = T_{sat} =$ **330 F**

$$x = \frac{v - v_f}{v_{fg}} = \frac{1.25 - 0.01776}{4.2938} = \mathbf{0.287}$$

$$u = u_f + x\, u_{fg} = 300.5 + 0.287 \times 805.68 = \mathbf{531.73\ Btu/lbm}$$

$$h = h_f + x\, h_{fg} = 300.84 + 0.287 \times 887.52 = \mathbf{555.56\ Btu/lbm}$$

b) Table F.7.1 : $T < T_{triple\ point}$ => F.7.4: $P > P_{sat}$ so **compressed solid**

$v \cong v_i =$ **0.01745 ft^3/lbm ;** $u \cong u_i =$ **-149.31 Btu/lbm** $h \cong h_i =$ **-149.31 kJ/kg**

approximate compressed solid with saturated solid properties at same T.

c) Table F.7.1: $u < u_g = 171.28$ Btu/lbm => **L+V mixture**, P = **138.93 psia**

$$x = \frac{u - u_f}{u_{fg}} = \frac{140 - 108.51}{62.77} = \mathbf{0.502},$$

$$v = v_f + x\, v_{fg} = 0.01387 + 0.502 \times 0.3278 = \mathbf{0.1784\ ft^3/lbm},$$

$$h = h_f + x\, h_{fg} = 108.86 + 0.502 \times 71.19 = \mathbf{144.6\ Btu/lbm}$$

d) Table F.9.1: $h > h_g$ => **superheated vapor,** look in F.9.2 between 180 and 200 F

$$T = 180 + (200 - 180)\frac{131 - 129.1}{133.03 - 129.1} = 180 + 20 \times 0.48346 = \mathbf{189.7\ F}$$

$$v = 0.3497 + (0.3661 - 0.3497) \times 0.48346 = \mathbf{0.35763\ ft^3/lbm}$$

$$u = h - Pv = 131 - 200 \times 0.35763 \times 144 / 778 = \mathbf{117.76\ Btu/lbm}$$

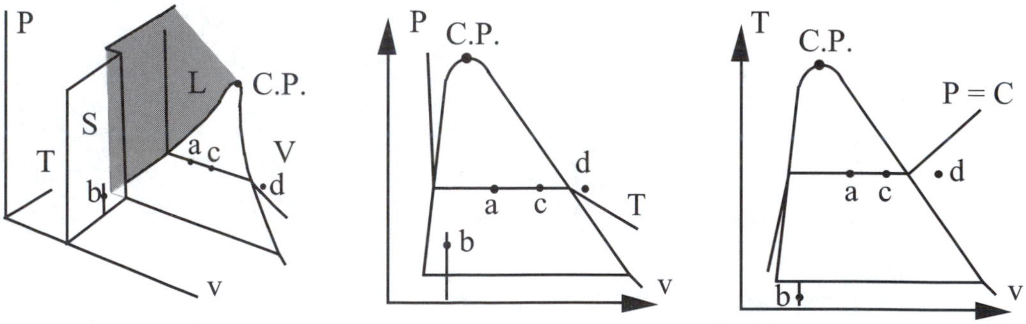

5.3E A two step process for a control mass

A piston/cylinder device contains 0.5 lbm water at 150 psia, 600 F. There is a constant force on the piston with stops mounted so a minimum volume is 1 ft^3. It is now cooled to 250 F. Find the final water volume, the work and heat transfer in the overall process and show the process in P-v and T-v diagrams.

Solution:

C.V. Water, this is a control mass. Process determined by device behavior.

Energy Eq.: $\qquad U_2 - U_1 = m(u_2 - u_1) = {}_1Q_2 - {}_1W_2$

State 1: (P, T) Table F.7.1: $v_1 = 4.111$ ft^3/lbm, $u_1 = 1211.58$ Btu/lbm

State 2: (T, ?)

Process: $\qquad P = \text{constant} = F/A = P_1 \quad \text{if} \quad V > V_{min}$

$\qquad\qquad V = \text{constant} = V_{1a} = V_{min} \quad \text{if} \quad P < P_1$

The only possible P-v combination for this system is shown in the diagram so both state 1 and 2 must be on the two lines. For state 2 we need to know if it is on the horizontal P line segment or the vertical v segment. Let us check state 1a:

State 1a: $\qquad P,\ v_{1a} = V_{min}/m = \frac{1}{0.5} = 2$ ft^3/lbm $\qquad$ from Table F.7.1 we get

$\qquad v_f < v_{1a} < v_g$ so two-phase and $T_{1a} = T_{sat} = 358$ F

We see that $T_2 < T_{1a}$ and state 2 must have $v_2 = v_{1a} = v_{min} = 2$ ft^3/lbm.

$$x_2 = \frac{v_2 - v_f}{v_{fg}} = \frac{2 - 0.017}{13.8077} = 0.1436$$

$$u_2 = u_f + x_2\, u_{fg} = 218.48 + x_2\ 869.41 = 343.33 \text{ Btu/lbm}$$

The work is the area under the process curve in the P-V diagram

$${}_1W_2 = m\ {}_1w_2 = \int_1^2 P\, dV = m \int_1^2 P\, dv = mP_1\,(v_{1a} - v_1) + 0$$

$$= 0.5 \text{ lbm} \times 150 \text{ psia} \times (2 - 4.111) \text{ ft}^3\text{/lbm}$$

$$= -0.5 \times 150 \times (2 - 4.111) \text{ lbf-ft-ft}^2\text{/in}^2 = -22\ 799 \text{ lbf-ft} = \mathbf{-29.3\ Btu}$$

Remember 1 ft^2/in^2 = 144. Now the heat transfer is found from the energy equation

$${}_1Q_2 = m(u_2 - u_1) + {}_1W_2 = 0.5\,(343.33 - 1211.58) - 29.3 = \mathbf{-463.4\ Btu}$$

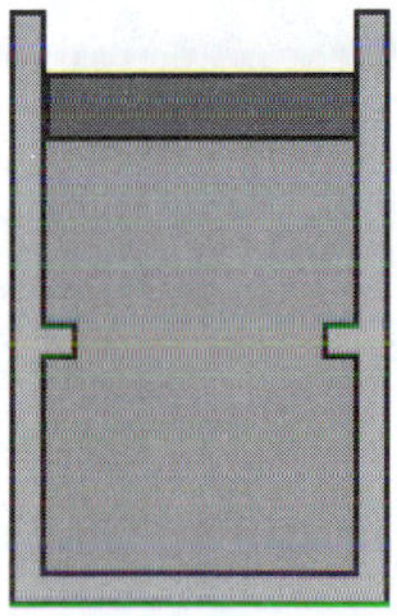

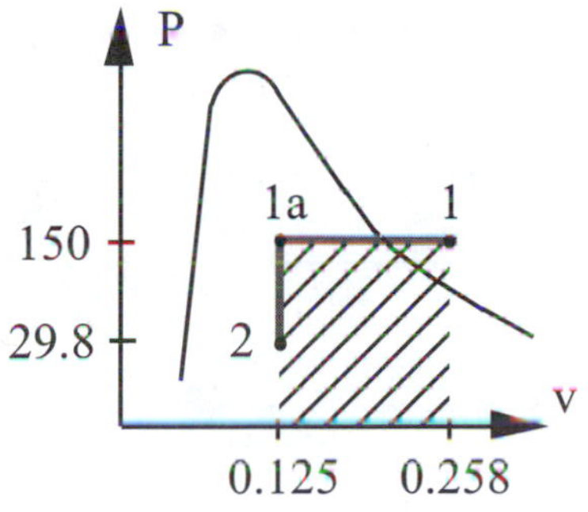

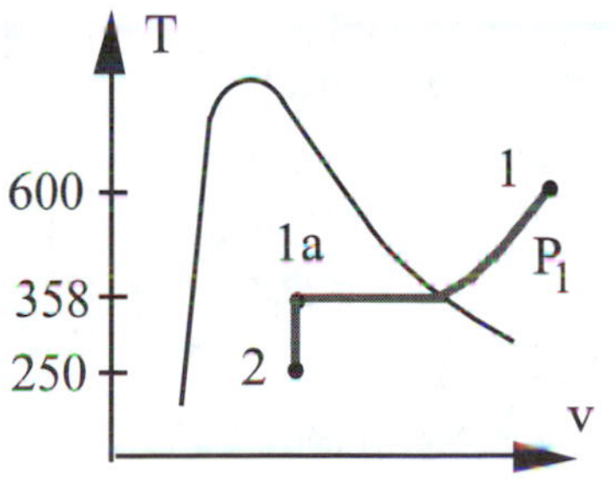

5.7E Heat exchange between solid and gas

A silicon wafer of volume 6 in^3 is heat treated at 500 F and now cooled in a 8-ft^3 closed air box initially at 77 F, 14.7 psia. The box itself consists of 4 lbm stainless 304 steel which has the same temperature as the air. Assuming no heat transfer with the surroundings, what is the final uniform temperature of all the masses? What is the final air pressure?

Solution:

C.V. The rigid steel box, the air and the silicon wafer so this is a control mass.

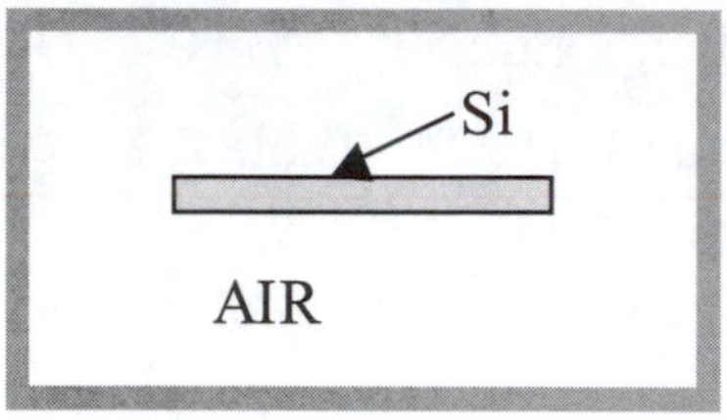

Process Eq: $V = \text{constant}\ ;\ {}_1W_2 = 0$ and ${}_1Q_2 = 0$

Energy Eq.: $U_2 - U_1 = {}_1Q_2 - {}_1W_2 = 0 - 0 = 0$

$$U_2 - U_1 = m_{steel}(u_2 - u_1)_{steel} + m_{air}(u_2 - u_1)_{air} + m_{si}(u_2 - u_1)_{si}$$

Properties from Tables F.2:

$$C_{steel} = 0.11 \text{ Btu/lbm R},\ C_{si} = 0.167 \text{ Btu/lbm R},\ \rho_{si} = 145.5 \text{ lbm/ft}^3$$

$$m_{si} = V\rho = 6 \times 12^{-3} \times 145.5 = 0.505 \text{ lbm}$$

Table F.4: $C_{V\,air} = 0.171$ Btu/lbm R, $R = 53.34$ ft-lbf/lbm R

Ideal gas: $$m_{air} = \frac{PV}{RT} = \frac{14.7 \text{ psia } (8 - 6 \times 12^{-3}) \text{ ft}^3 \times 144 \text{ (in/ft)}^2}{53.34 \times (77 + 459.7) \text{ ft-lbf/lbm}} = 0.5913 \text{ lbm}$$

For solids and liquids Eq.5.17, $\Delta u \cong C\,\Delta T$, similar to a gas Eq.5.20: $\Delta u \cong C_V\,\Delta T$,

The energy equation for the C.V. becomes

$$m_{steel}C_{steel}(T_2 - T_{1,steel}) + m_{si}C_{si}(T_2 - T_{1,si}) + m_{air}C_{V\,air}(T_2 - T_{1,air}) = 0$$

Substitute values

$$4 \times 0.11\,(T_2 - 77) + 0.505 \times 0.167\,(T_2 - 500) + 0.5913 \times 0.171\,(T_2 - 77) = 0$$

$$0.625447\,T_2 - 33.8 - 42.1675 - 7.78565 = 0$$

Solve for T_2: $\Rightarrow T_2 = \mathbf{134\ F}$

From the process and the energy equations we have state 2: (T_2, $v_2 = v_1$). Using then the ideal gas law, PV = mRT, for states 1 and 2 gives

$$P_2 = P_1 \frac{V_1}{V_2}\frac{mRT_2}{mRT_1} = 14.7\,\frac{459.7 + 134}{459.7 + 77} = \mathbf{16.3\ psia}$$

CHAPTER 6
STUDY PROBLEMS

ENERGY EQUATION FOR A CONTROL VOLUME

- **Conservation of mass, continuity equation**
- **Conservation of energy, the first law**
- **Steady state processes and devices**
- **The transient process**

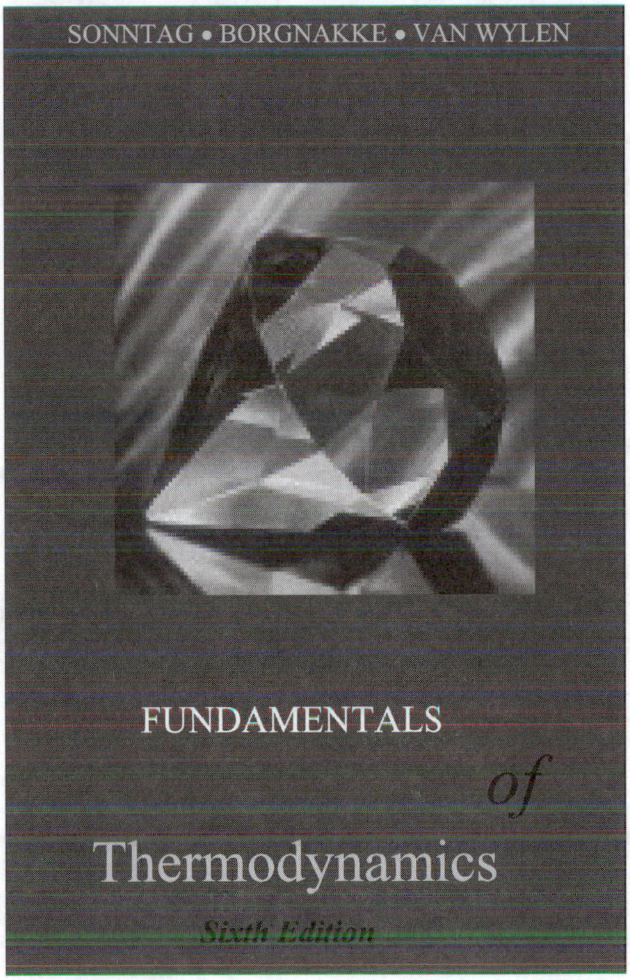

6.1 Steady pipe flow

Assume we have a low velocity flow in a long pipe of radius R. For such a flow the velocity is maximum at the centerline and varies as a parabola

$$V(r) = V_{max}\left[\ 1 - \left(\frac{r}{R}\right)^2\ \right]$$

We want to find the total mass flow rate and the average velocity.

Solution:

The mass flow rate is summed up over the area as in Eq.6.3

$$\begin{aligned}\dot{m} &= \rho A\mathbf{V}_{avg} = A\mathbf{V}_{avg} / v = \int \rho\mathbf{V}\, dA = \int \rho\mathbf{V}\, 2\pi\, r\, dr \\ &= 2\pi\rho \int_0^R \mathbf{V}_{max}\, r\left[\ 1 - \left(\frac{r}{R}\right)^2\ \right] dr \\ &= 2\pi\rho\, \mathbf{V}_{max}\left[\frac{1}{2} r^2 - \frac{1}{R^2}\frac{1}{4} r^4\right]_0^R \\ &= \frac{1}{2}\pi\rho\, \mathbf{V}_{max}\, R^2\end{aligned}$$

The cross sectional area is $A = \pi R^2$ so the average velocity becomes

$$\mathbf{V}_{avg} = \dot{m} / \rho\pi R^2 = \frac{1}{2}\mathbf{V}_{max}$$

significant lower than the maximum (centerline) velocity. The flow area is small (r is small) where the velocity is large and the area is large (r close to R) where the velocity is small which explains the low average.

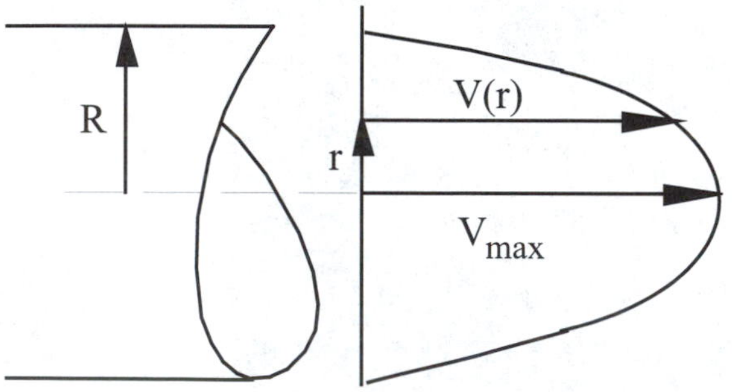

6.2 Steady liquid flow through a pump and nozzle

A harbor fire and rescue boat has a water cannon which flows 5 kg/s water, 15°C, in a 5 cm diameter pipe to a nozzle with an exit flow diameter of 1.5 cm. We would like to know the velocities in and out of the nozzle and the total pump work delivered to the water assuming pump inlet pressure is equal to the nozzle exit pressure of 101 kPa.

Solution:
Take as a control volume the nozzle. This is steady state, single flow in and out with no heat transfer (water is at ambient temperature).

Continuity Eq.6.11, 6.3: $\dot{m} = \rho_i A_i \mathbf{V}_i = \rho_e A_e \mathbf{V}_e$

Energy Eq.6.13: $(h + \frac{1}{2}\mathbf{V}^2 + gZ)_i = (h + \frac{1}{2}\mathbf{V}^2 + gZ)_e$

Since it is liquid water the density (specific volume) is constant $\rho = \rho_i = \rho_e = 1/v$ so from table B.1.1: $\rho = 1 / 0.001001 = 999$ kg/m^3. From the mass flow rate and the continuity equation we can get the velocities as

$$\mathbf{V}_i = \dot{m} / \rho A_i = \frac{5 \text{ kg/s}}{999 \text{ (kg/m}^3) (\pi/4)\, 0.05^2 \text{ m}^2} = \mathbf{2.55 \text{ m/s}}$$

$$\mathbf{V}_e = \mathbf{V}_i\, A_i / A_e = \mathbf{V}_i \left(\frac{D_i}{D_e}\right)^2 = 2.55 \left(\frac{0.05}{0.015}\right)^2 = \mathbf{28.3 \text{ m/s}}$$

C.V. The pump, pipe and nozzle. Assume the pump has an inlet from a storage tank or from the harbor surface water with zero kinetic energy, no heat transfer and $Z_i = Z_e$.

Energy Eq. 6.13: $h_i + 0 + gZ_i = h_e + \frac{1}{2}\mathbf{V}_e^2 + gZ_e + w$

and since (P, T) of the liquid in and out is the same ($h_i = h_e$) the work becomes

$$\dot{W}_{pump} = \dot{m}w = \dot{m}[h_i - h_e + g(Z_i - Z_e) - \tfrac{1}{2}\mathbf{V}_e^2\,] = -\,\dot{m}\,\tfrac{1}{2}\mathbf{V}_e^2$$

$$= -\,5 \text{ kg/s} \times 0.5 \times 28.3^2 \text{ m}^2/\text{s}^2 = -\,\mathbf{2002 \text{ W} = -2.0 \text{ kW}}$$

having converted (kg/s)(m^2/s^2) to W and then to kW by division with 1000.

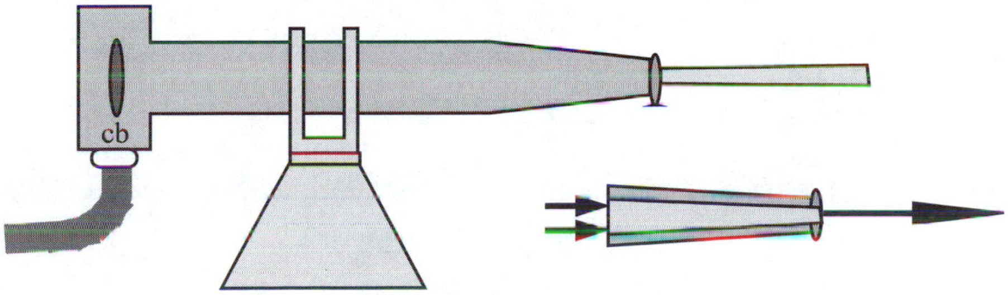

6.3 A stagnation flow problem

A horizontal flow at 40 m/s, 300 K and 100 kPa approaches the front surface of a blunt body (this could be the front of a car, a submarine or the wing of an airplane). At the dividing streamline the flow comes to a stop on the stagnation point on the surface, away from this line the flow goes around the body. We would like to know the temperature if the flow is liquid water and if it is air.

Solution:

C.V. Small section around the dividing streamline, steady state, single inlet and exit flow. The flow is horizontal so it has a constant elevation (Z = 0) and we will assume that there is no heat transfer.

Energy Eq.6.13: $$0 + h_1 + \frac{1}{2}\mathbf{V}_1^2 + 0 = h_2 + 0 + 0 + 0$$

Here we also used $\mathbf{V}_2 \cong 0$ and this h is called the stagnation enthalpy

$$h_2 = h_1 + \frac{1}{2}\mathbf{V}_1^2$$

Let us assume we can use a constant heat capacity from table A.4 or A.5 then,

$$T_2 = T_1 + \frac{1}{2}\mathbf{V}_1^2 / C_p$$

For the liquid water $C_p = 4.18$ kJ/kg K = 4180 J/kg K

$$T_2 = 300 + \frac{1}{2} 40^2 / 4180 = \mathbf{300.19\ K}$$

For the air $C_p = 1.004$ kJ/kg K = 1004 J/kg K

$$T_2 = 300 + \frac{1}{2} 40^2 / 1004 = \mathbf{300.8\ K}$$

The temperature increase is not very large in any of the cases. The kinetic energy will be much higher for higher velocities like a meteorite falling to earth where the T is so high that the rock melts and evaporates (due to oxygen it burns up).

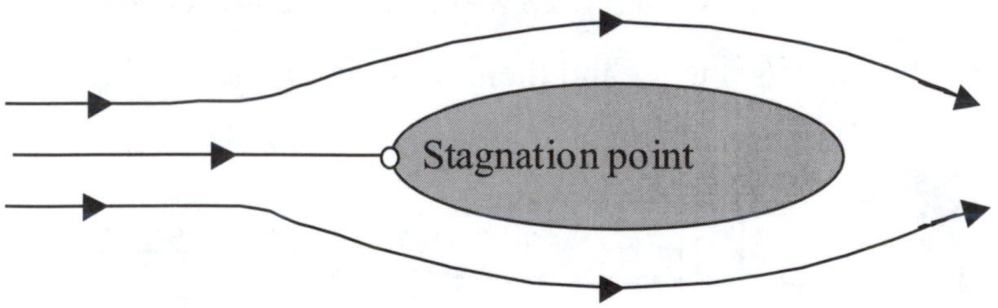

6.4 Air flow in a jet engine exit nozzle

Air leaving the turbine section in a jet engine at 1100 K, 450 kPa with a velocity of 200 m/s enters a nozzle with an exit state of 100 kPa and 780 K. Find the exit velocity and the ratio of the exit area to the nozzle inlet area A_{ex}/A_{in}.

Solution:

C.V. around the nozzle section, this is steady state, single flow with no heat transfer and no work, also same elevation $Z_{in} = Z_{ex}$.

Continuity Eq.: $\dot{m}_{in} = \dot{m}_{ex} = \dot{m} = A_{in}\mathbf{V}_{in}/v_{in} = A_{ex}\mathbf{V}_{ex}/v_{ex}$

Energy Eq.6.12: $\dot{m}(h_{in} + \frac{1}{2}\mathbf{V}_{in}^2) = \dot{m}(h_{ex} + \frac{1}{2}\mathbf{V}_{ex}^2)$

State properties from Table A.7: $h_{in} = 1161.18$ kJ/kg, $h_{ex} = 800.28$ kJ/kg.

From the energy equation solve for the exit kinetic energy, notice conversion from kJ to J for the enthalpy terms

$$\frac{1}{2}\mathbf{V}_{ex}^2 = \frac{1}{2}\mathbf{V}_{in}^2 + h_{in} - h_{ex} = \frac{1}{2}\,200^2 + (1161.18 - 800.28)\,1000$$

$$= 20\,000 + 360\,900 = 380\,900\ \mathrm{m^2/s^2}$$

$$\mathbf{V}_{ex} = \sqrt{2 \times 380\,900} = \mathbf{872.8\ m/s}$$

From the continuity equation and ideal gas law to get $v = RT/P$ we have

$$A_{ex} / A_{in} = \frac{\mathbf{V}_{in}}{\mathbf{V}_{ex}} \times \frac{T_{ex}}{T_{in}} \times \frac{P_{in}}{P_{ex}} = \frac{200}{872.8} \times \frac{780}{1100} \times \frac{450}{100} = \mathbf{0.731}$$

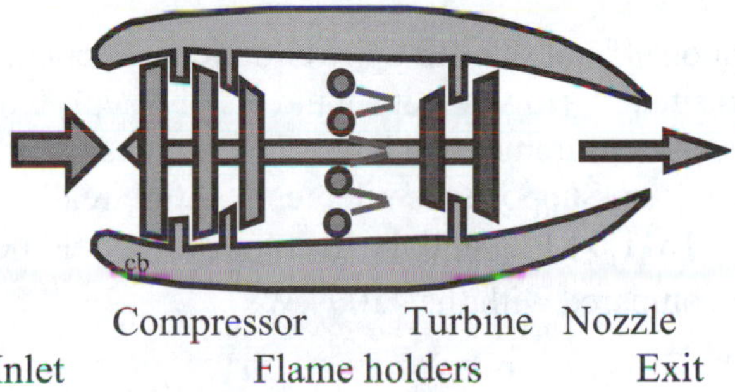

6.5 A multiple inlet flow turbine

A turbine in a gas turbine power plant receives 2 kg/s air at 3 MPa, 1500 K from one source and 5 kg/s air at 2 MPa, 1000 K from a different source. The exit state is 200 kPa, 600 K and it is measured that there is a total power out on the shaft of 4200 kW. We would like to know if this turbine has a significant heat loss.

Solution:

C.V. Turbine steady state, 2 inlets and 1 exit flow, neglect the kinetic and potential energy differences.

Continuity Eq.6.9: $\dot{m}_1 + \dot{m}_2 = \dot{m}_3 = 2 + 5 = 7$ kg/s

Energy Eq.6.10: $\dot{m}_1 h_1 + \dot{m}_2 h_2 + \dot{Q} = \dot{m}_3 h_3 + \dot{W}_T$

Table A.7: $h_1 = 1635.8$ kJ/kg,

$h_2 = 1046.22$ kJ/kg

$h_3 = 607.32$ kJ/kg

$\dot{W}_T = 4200$ kW

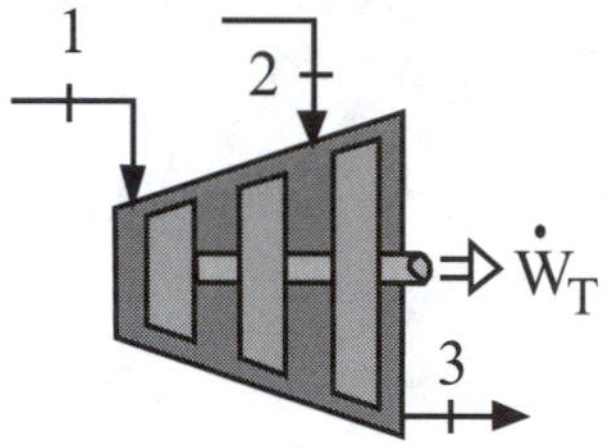

Solve for the heat transfer in the energy equation

$$\dot{Q} = \dot{m}_3 h_3 + \dot{W}_T - \dot{m}_1 h_1 - \dot{m}_2 h_2$$

$$= 7 \times 607.32 + 4200 - 5 \times 1046.22 - 2 \times 1635.8 = \mathbf{-\,51.5\ kW}$$

Comment:

So we have a loss of about 1% of the shaft power output. If economically feasible the turbine can be better insulated. However it should also be kept in mind that the state specifications and power measurement all have uncertainties, which also could explain the small, apparent heat loss. Suppose the inlet temperature really was only 1480 K then the inlet enthalpy $h_1 = 1611.6$ kJ/kg and the heat transfer would be found to –3.1 kW which indeed is small compared with the 4200 kW.

6.6 The condenser in a refrigerator

A condenser in a refrigerator receives a flow of R-12 at 1000 kPa, 60°C, which is cooled by kitchen air coming in at 20°C so the R-12 leaves as saturated liquid at 40°C and the air leaves at 30°C. If the condenser should remove a total of 1 kW from the R-12 line, find the mass flow rate of both fluids.

Solution:

CV Condenser. This is a two-fluid heat exchanger with R-12 inside a metal line usually with fins or plates attached over which flows air without being enclosed. A fan drives the flow of air if the condenser is located at the bottom or it is driven by natural convection when the condenser is located on the back side of the refrigerator. We neglect any work.

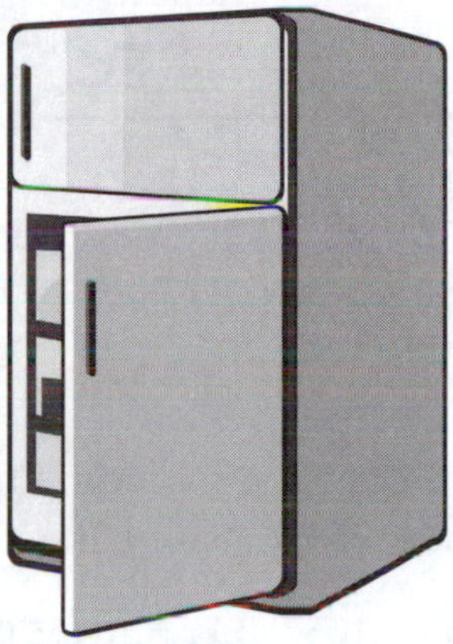

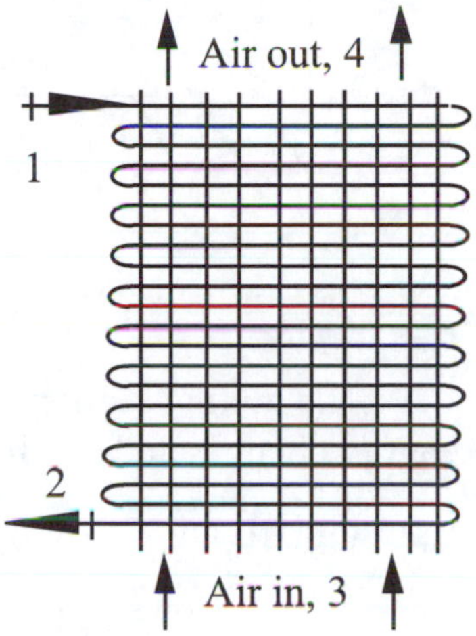

Continuity equations: $\dot{m}_1 = \dot{m}_2 = \dot{m}_{R\text{-}12}$ and $\dot{m}_3 = \dot{m}_4 = \dot{m}_{air}$

Energy Eq.: $\dot{m}_{R\text{-}12}h_1 + \dot{m}_{air}h_3 = \dot{m}_{R\text{-}12}h_2 + \dot{m}_{air}h_4$

R-12 from Table B.3: $h_1 = 217.97$ kJ/kg, $h_2 = 74.59$ kJ/kg

Air from Table A.5: $h_4 - h_3 = C_p(T_4 - T_3) = 1.004\ (30 - 20) = 10.04$ kJ/kg

Let us rearrange the energy equation to show the terms that would appear if we take a C.V. around each flow:

$$\dot{m}_{air}(h_4 - h_3) = \dot{Q}_{air} = -\dot{Q}_{R\text{-}12} = -\dot{m}_{R\text{-}12}(h_2 - h_1) = 1 \text{ kW}$$

It now follows that the flow rates are:

$$\dot{m}_{R\text{-}12} = -\dot{Q}_{R\text{-}12} / (h_1 - h_2) = \frac{1}{217.97 - 74.59}\frac{\text{kW}}{\text{kJ/kg}} = \mathbf{0.007\ kg/s}$$

$$\dot{m}_{air} = \dot{Q}_{air} / (h_4 - h_3) = 1 \text{ kW} / 10.04 \text{ kJ/kg} = \mathbf{0.1\ kg/s}$$

Notice: We do not use heat capacity for the R-12 since it changes phase.

6.7 Mixing of multiple flows

An underground garage and utility rooms has 4 fans blowing air into a ventilation duct at steady state conditions. We would like to know the exit temperature and velocity. How much power is needed to create the kinetic energy in the exit flow?

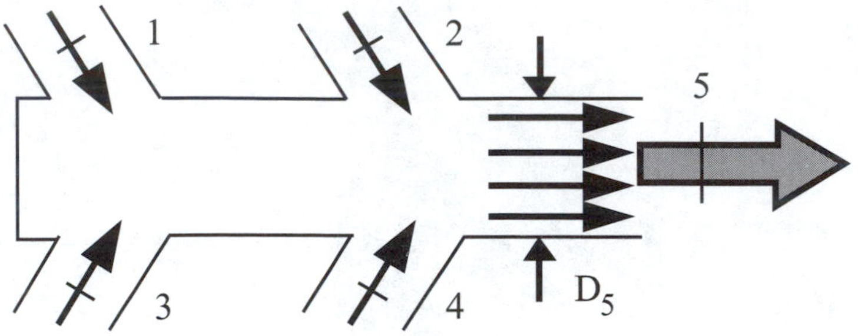

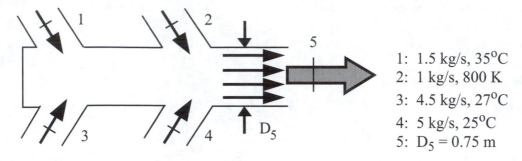

Mixing section

Solution:

C.V. The mixing section of the duct not including the fans. We will also neglect kinetic energy changes and any heat transfer (assume the section we look at is short).

Continuity Eq. 6.9: $\dot{m}_1 + \dot{m}_2 + \dot{m}_3 + \dot{m}_4 = \dot{m}_5 = \rho_5 A_5 V_5$

Energy Eq.6.10: $\dot{m}_1 h_1 + \dot{m}_2 h_2 + \dot{m}_3 h_3 + \dot{m}_4 h_4 = \dot{m}_5 h_5$

Since the temperatures are modest we use constant heat capacity, $h = h_o + C_p(T - T_o)$, and the h_o and T_o terms cancel out in the energy equation. To see this, multiply the continuity equation with $h_o - C_p T_o$ and subtract from the energy equation. We can solve for T_5 after dividing C_p out

$$T_5 = \frac{\dot{m}_1}{\dot{m}_5} T_1 + \frac{\dot{m}_2}{\dot{m}_5} T_2 + \frac{\dot{m}_3}{\dot{m}_5} T_3 + \frac{\dot{m}_4}{\dot{m}_5} T_4$$

$$= \frac{1.5}{12} 35 + \frac{1}{12} (800 - 273) + \frac{4.5}{12} 27 + \frac{5}{12} 25 = \mathbf{68.8^oC}$$

For the exit state 5 we then get:

$$\rho_5 = \frac{1}{v_5} = \frac{P}{RT_5} = \frac{101}{0.287\,(68.8 + 273.15)} = 1.032 \text{ kg/m}^3$$

$$A_5 = \frac{1}{4} \pi\, D^2 = \frac{1}{4} \pi\, 0.75^2 = 0.4418 \text{ m}^2$$

$$V_5 = \dot{m}_5 / (\rho_5 A_5) = \frac{12}{1.032 \times 0.4418} = \mathbf{26.32 \text{ m/s}}$$

The total kinetic energy flow out is:

$$\dot{W}_{\text{kin flow}} = \dot{m}_5 \frac{1}{2} V_5^2 = 12 \times 26.32\,^2/2000 = \mathbf{4.16 \text{ kW}}$$

6.8 A steady flow mixing chamber (feed water heater)

A flow at state 1 of 10 kg/s water at 75 kPa, 40°C is mixed with steam flowing at state 2 which is at 100 kPa, 150°C. Which mass flow rate at state 2 will produce an exit flow of saturated liquid at 75 kPa with no external heat transfer?

Solution:

C.V Mixing chamber and valve. Steady state no heat transfer or work terms.

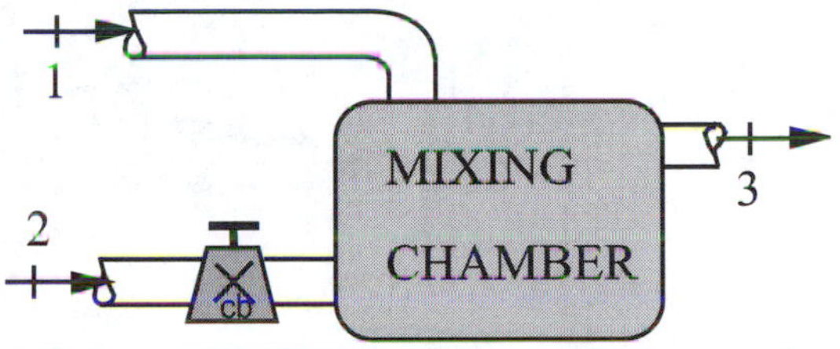

Continuity Eq.6.9: $\dot{m}_1 + \dot{m}_2 = \dot{m}_3$

Energy Eq.6.10: $\dot{m}_1 h_1 + \dot{m}_2 h_2 = \dot{m}_3 h_3 = (\dot{m}_1 + \dot{m}_2) h_3$

Properties Table B.1.1: $h_1 = 167.54$ kJ/kg ; assume saturated liquid same T

Table B.1.3: $h_2 = 2776.38$ kJ/kg superheated vapor

Table B.1.2: $h_3 = 384.36$ kJ/kg h_f at 75 kPa

Solve for $\dot{m}_2$ from the energy equation

$$\dot{m}_2 = \dot{m}_1 \times \frac{h_1 - h_3}{h_3 - h_2} = 10 \times \frac{167.54 - 384.36}{384.36 - 2776.38} = \mathbf{0.906\ kg/s}$$

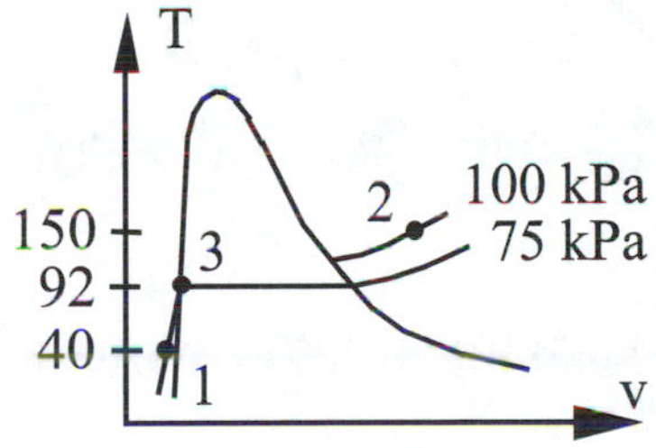

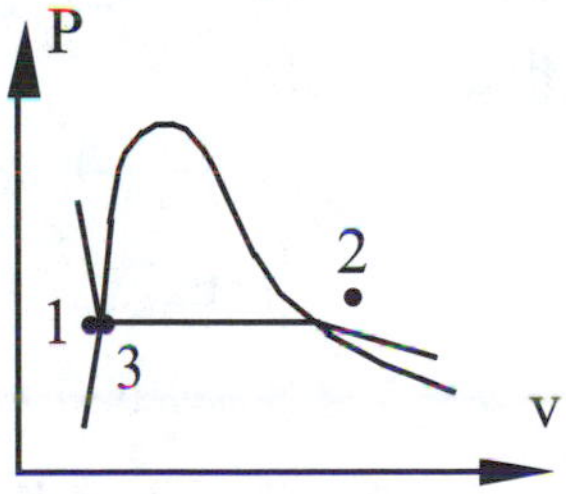

Remark: This device is used in a steam power plant to heat the feed water state 1 with extraction steam from the turbine, state 2, to produce a warmer liquid flow, state 3.

6.9 A transient filling process

A rigid tank of volume 2 m^3 contains air at ambient 100 kPa, 290 K. It is slowly filled with air from a supply line at 1000 kPa and 300 K. After a long time the flow stops and the final temperature in the tank equals the ambient 290 K. How much mass was added to the tank and what was the heat transfer.

Solution:

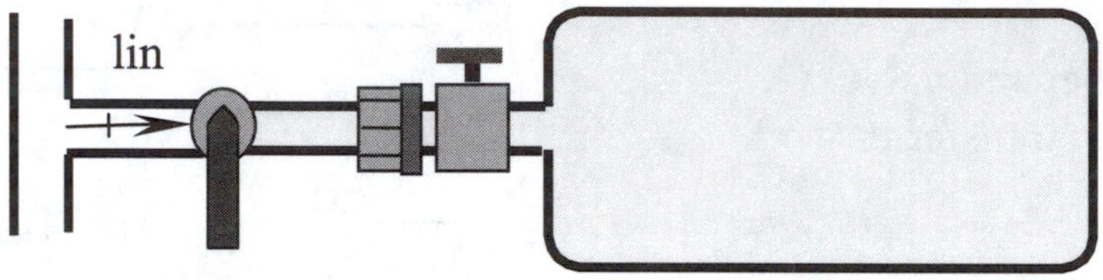

CV The tank and its content.

Continuity Eq.6.15: $m_2 - m_1 = m_i$

Energy Eq.6.16: $m_2u_2 - m_1u_1 = m_ih_i + {}_1Q_2 - {}_1W_2$

Process: $V = C \Rightarrow {}_1W_2 = \int P\, dV = 0$

State 1 (ideal gas): $m_1 = P_1V_1 / RT_1 = 100 \times 2 / (0.287 \times 290) = 2.403$ kg

When the flow stops we must have line pressure inside the tank.

State 2: $(T_2, P) \Rightarrow m_2 = P_2V_2 / RT_2 = 1000 \times 2 / (0.287 \times 290) = 24.03$ kg

From the continuity equation

$$m_i = m_2 - m_1 = \mathbf{21.627\ kg}$$

Now in the energy equation, $u_2 = u_1$ so we get

$$m_2u_2 - m_1u_1 = m_iu_2 = m_ih_i + {}_1Q_2 - 0$$

so solve for the heat transfer

$$
\begin{aligned}
{}_1Q_2 &= m_iu_2 - m_ih_i = m_i(u_2 - h_i) = m_i(u_2 - u_i - RT_i) \\
&= m_i[\ C_v(T_2 - T_i) - RT_i] \\
&= 21.627\ [\ 0.717(290 - 300) - 0.287 \times 300] \\
&= \mathbf{-2017\ kJ}
\end{aligned}
$$

Comment: Not only do we have to cool down from the line T, but we also have to remove the energy of the flow work that was coming in.

6.10 A transient exit flow process

Air in an insulated piston/cylinder setup is at 200 kPa, 360 K with a volume of 0.1 m^3. There is a constant force on the piston. The air runs out through a nozzle and the exit temperature is measured to 300 K. The process continues to a final air volume of 0.01 m^3. Find the exit velocity.

Solution:

C.V. Cylinder and nozzle. We assume a uniform pressure inside (neglect nozzle volume) of 200 kPa during the whole process and zero heat transfer since it is insulated.

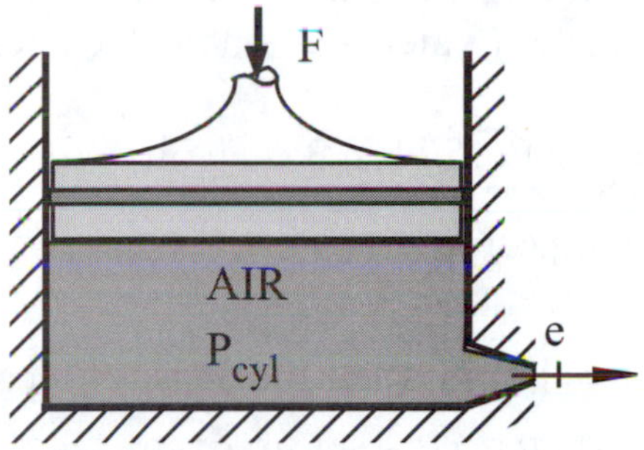

Continuity Eq.6.15: $m_2 - m_1 = -m_e$

Energy Eq.6.16: $m_2u_2 - m_1u_1 = -m_e(h_e + \frac{1}{2}\mathbf{V}_e^2) - {}_1W_2$

Process: $P = C \quad \Rightarrow \quad {}_1W_2 = \int P\, dV = P(V_2 - V_1)$

$$_1W_2 = P(V_2 - V_1) = 200 \text{ kPa } (0.01 - 0.1) \text{ m}^3 = -18 \text{ kJ}$$

Air is an ideal gas and the state inside does not change, $T_2 = T_1$ so

State 1: $(T_1, P) \Rightarrow \quad m_1 = P_1V_1 / RT_1 = 200 \times 0.1 / (0.287 \times 360) = 0.1936$ kg

State 2: $(T_2, P) \Rightarrow \quad m_2 = P_2V_2 / RT_2 = 200 \times 0.01 / (0.287 \times 360) = 0.0194$ kg

$$m_e = m_1 - m_2 = 0.1742 \text{ kg}$$

Now in the energy equation, $u_2 = u_1$ so we get

$$m_2u_2 - m_1u_1 = -m_eu_1 = -m_e(h_e + \tfrac{1}{2}\mathbf{V}_e^2) - {}_1W_2$$

and solve for the kinetic energy

$$\tfrac{1}{2}\mathbf{V}_e^2 = u_1 - h_e - {}_1W_2 / m_e = C_v(T_1 - T_e) - RT_e - {}_1W_2 / m_e$$

$$= 0.717\ (360 - 300) - 0.287 \times 300 + 18 / 0.1742$$

$$= 43.02 - 86.1 + 103.33 = 60.25 \text{ kJ/kg}$$

The velocity becomes

$$\mathbf{V}_e = \sqrt{2 \times 60.25 \times 1000} = \mathbf{347\ m/s}$$

6.2E Steady liquid flow through a pump and nozzle

A harbor fire and rescue boat has a water cannon which flows 10 lbm/s water, 60 F, in a 2 in. diameter pipe to a nozzle with an exit flow diameter of 0.5 in. We would like to know the velocities in and out of the nozzle and the total pump work delivered to the water assuming pump inlet pressure is equal to the nozzle exit pressure of 14.7 psia.

Solution:

Take as a control volume the nozzle. This is steady state, single flow in and out with no heat transfer (water is at ambient temperature).

Continuity Eq.6.11, 6.3: $\dot{m} = \rho_i A_i \mathbf{V}_i = \rho_e A_e \mathbf{V}_e$

Energy Eq.6.13: $(h + \frac{1}{2}\mathbf{V}^2 + gZ)_i = (h + \frac{1}{2}\mathbf{V}^2 + gZ)_e$

Since it is liquid water the density (specific volume) is constant $\rho = \rho_i = \rho_e = 1/v$ so from table F.7.1: $\rho = 1 / 0.01603 = 62.383$ lbm/ft^3. From the mass flow rate and the continuity equation we can get the velocities as

$$\mathbf{V}_i = \dot{m} / \rho A_i = \frac{10 \text{ lbm/s}}{62.383 \text{ (lbm/ft}^3) (\pi/4) (2/12)^2 \text{ ft}^2} = \mathbf{7.35 \text{ ft/s}}$$

$$\mathbf{V}_e = \mathbf{V}_i \, A_i / A_e = \mathbf{V}_i \left(\frac{D_i}{D_e}\right)^2 = 7.35 \left(\frac{2}{0.5}\right)^2 = \mathbf{117.6 \text{ ft/s}}$$

C.V. The pump, pipe and nozzle. Assume the pump has an inlet from a storage tank or from the harbor surface water with zero kinetic energy, no heat transfer and $Z_i = Z_e$.

Energy Eq. 6.13: $h_i + 0 + gZ_i = h_e + \frac{1}{2}\mathbf{V}_e^2 + gZ_e + w$

and since (P, T) of the liquid in and out is the same ($h_i = h_e$) the work becomes

$$\dot{W}_{pump} = \dot{m}w = \dot{m}[h_i - h_e + g(Z_i - Z_e) - \tfrac{1}{2}\mathbf{V}_e^2] = - \dot{m} \tfrac{1}{2}\mathbf{V}_e^2$$

$$= - 10 \text{ lbm/s} \times 0.5 \times 117.6^2 \text{ ft}^2/\text{s}^2 = - \mathbf{2.76 \text{ Btu/s}}$$

having converted (lbm/s)(ft^2/s^2) to Btu/s by multiplication with 3.9941×10^{-5} (or div by 25 037) see table A.1 under specific kinetic energy.

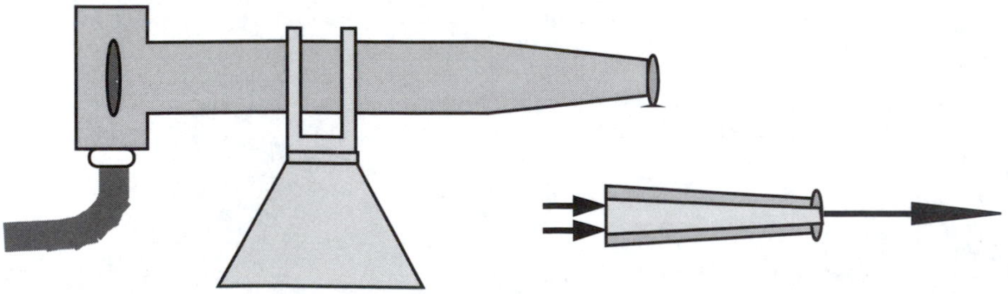

6.4E Air flow in a jet engine exit nozzle

Air leaving the turbine section in a jet engine at 2000 R, 70 psia with a velocity of 600 ft/s enters a nozzle with an exit state of 14.7 psia and 1400 R. Find the exit velocity and the ratio of the exit area to the nozzle inlet area A_{ex}/A_{in}.

Solution:

C.V. around the nozzle section, this is steady state, single flow with no heat transfer and no work, also same elevation $Z_{in} = Z_{ex}$.

Continuity Eq.: $\dot{m}_{in} = \dot{m}_{ex} = \dot{m} = A_{in}\mathbf{V}_{in}/v_{in} = A_{ex}\mathbf{V}_{ex}/v_{ex}$

Energy Eq.6.12: $\dot{m}(h_{in} + \frac{1}{2}\mathbf{V}_{in}^2) = \dot{m}(h_{ex} + \frac{1}{2}\mathbf{V}_{ex}^2)$

State properties from Table F.5: $h_{in} = 504.755$ Btu/lbm, $h_{ex} = 343.016$ Btu/lbm.

From the energy equation solve for the exit kinetic energy, notice conversion from Btu/lbm to ft^2/s^2 by multiplication with 25 037 (see Table A.1) for the enthalpy terms

$$\frac{1}{2}\mathbf{V}_{ex}^2 = \frac{1}{2}\mathbf{V}_{in}^2 + h_{in} - h_{ex} = \frac{1}{2}600^2 + (504.755 - 343.016) \times 25\,037$$

$$= 180\,000 + 4\,049\,459 = 4\,229\,459 \text{ ft}^2/\text{s}^2$$

$$\mathbf{V}_{ex} = \sqrt{2 \times 4\,049\,459} = \mathbf{2908 \text{ ft/s}}$$

From the continuity equation and ideal gas law to get $v = RT/P$ we have

$$A_{ex} / A_{in} = \frac{\mathbf{V}_{in}}{\mathbf{V}_{ex}} \times \frac{T_{ex}}{T_{in}} \times \frac{P_{in}}{P_{ex}} = \frac{600}{2908} \times \frac{1400}{2000} \times \frac{70}{14.7} = \mathbf{0.688}$$

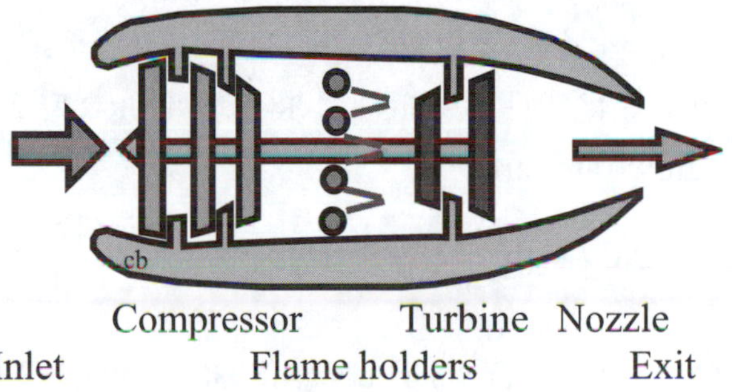

6.6E The condenser in a refrigerator

A condenser in a refrigerator receives a flow of R-22 at 150 psia, 140 F, which is cooled by kitchen air coming in at 68 F so the R-22 leaves as saturated liquid at 100 F and the air leaves at 88 F. If the condenser should remove a total of 1 Btu/s from the R-22 line, find the mass flow rate of both fluids.

Solution:

CV Condenser. This is a two-fluid heat exchanger with R-22 inside a metal line usually with fins or plates attached over which flows air without being enclosed. A fan drives the flow of air if the condenser is located at the bottom or it is driven by natural convection when the condenser is located on the back side of the refrigerator. We neglect any work.

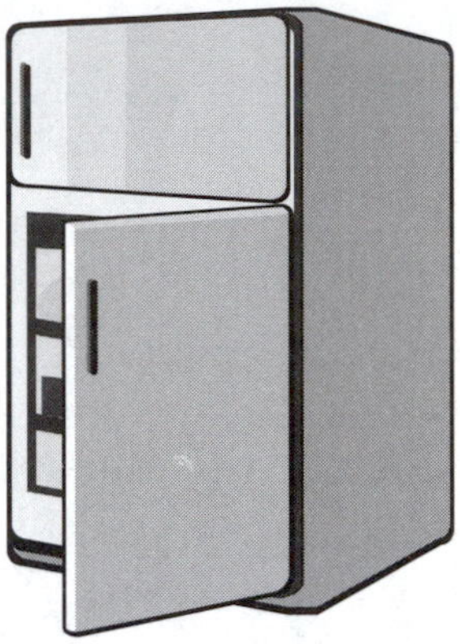

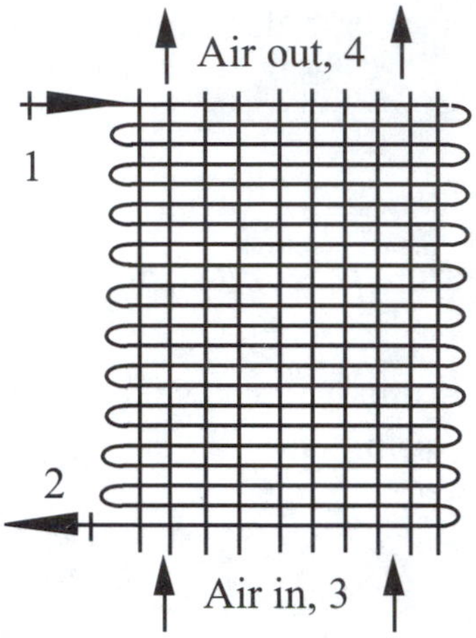

Continuity equations: $\dot{m}_1 = \dot{m}_2 = \dot{m}_{R\text{-}22}$ and $\dot{m}_3 = \dot{m}_4 = \dot{m}_{air}$

Energy Eq.: $\dot{m}_{R\text{-}12}h_1 + \dot{m}_{air}h_3 = \dot{m}_{R\text{-}22}h_2 + \dot{m}_{air}h_4$

R-12 from Table F.9: $h_1 = 123.18$ Btu/lbm, $h_2 = 39.27$ Btu/lbm

Air from Table F.4: $h_4 - h_3 = C_p(T_4 - T_3) = 0.24\ (88 - 68) = 4.8$ Btu/lbm

Let us rearrange the energy equation to show the terms that would appear if we take a C.V. around each flow:

$$\dot{m}_{air}(h_4 - h_3) = \dot{Q}_{air} = -\dot{Q}_{R\text{-}22} = -\dot{m}_{R\text{-}22}(h_2 - h_1) = 1 \text{ Btu/s}$$

It now follows that the flow rates are:

$$\dot{m}_{R\text{-}22} = -\dot{Q}_{R\text{-}22} / (h_1 - h_2) = \frac{1}{123.18 - 39.27}\ \frac{\text{Btu/s}}{\text{Btu/lbm}} = \mathbf{0.0119\ lbm/s}$$

$$\dot{m}_{air} = \dot{Q}_{air} / (h_4 - h_3) = 1 \text{ Btu/s} / 4.8 \text{ Btu/lbm} = \mathbf{0.208\ lbm/s}$$

Notice: We do not use heat capacity for the R-22 since it changes phase.

6.9E A transient filling process

A rigid tank of volume 70 ft^3 contains air at ambient 14.7 psia, 520 R. It is slowly filled with air from a supply line at 150 psia and 540 R. After a long time the flow stops and the final temperature in the tank equals the ambient 520 R. How much mass was added to the tank and what was the heat transfer.

Solution:

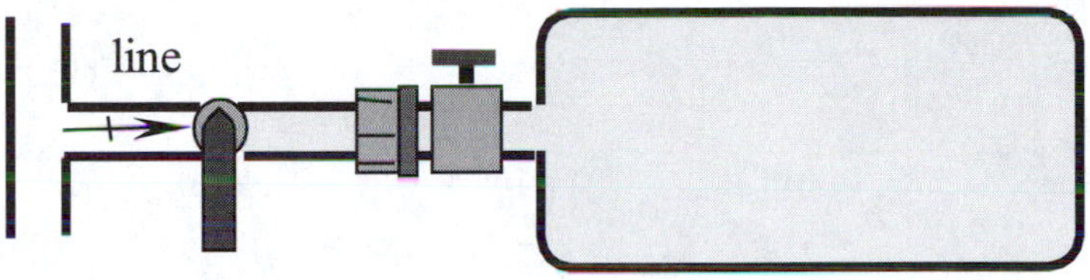

CV The tank and its content.

Continuity Eq.6.15: $m_2 - m_1 = m_i$

Energy Eq.6.16: $m_2u_2 - m_1u_1 = m_ih_i + {}_1Q_2 - {}_1W_2$

Process: $V = C \Rightarrow {}_1W_2 = \int P\, dV = 0$

State 1 (ideal gas): $m_1 = \frac{P_1V_1}{R\,T_1} = \frac{14.7 \text{ lbf/in}^2 \times 70 \text{ ft}^3 \times 144 \text{ (in/ft)}^2}{53.34 \text{ lbf-ft/lbm-R} \times 520 \text{ R}} = 5.342 \text{ lbm}$

When the flow stops we must have line pressure inside the tank.

State 2: $(T_2, P) \Rightarrow \quad m_2 = P_2V_2 / RT_2 = 150 \times 70 \times 144/ (53.34 \times 520) = 54.512$ lbm

From the continuity equation

$$m_i = m_2 - m_1 = \mathbf{49.17\ lbm}$$

Now in the energy equation, $u_2 = u_1$ so we get

$$m_2u_2 - m_1u_1 = m_iu_2 = m_ih_i + {}_1Q_2 - 0$$

so solve for the heat transfer (convert 778 lbf-ft = 1 Btu)

$$
\begin{aligned}
{}_1Q_2 &= m_iu_2 - m_ih_i = m_i(u_2 - h_i) = m_i(u_2 - u_i - RT_i) \\
&= m_i[\ C_v(T_2 - T_i) - RT_i] \\
&= 49.17\ [\ 0.171\ (520 - 540) - 53.34 \times 540 / 778] \\
&= \mathbf{-1988\ Btu}
\end{aligned}
$$

Comment: Not only do we have to cool down from the line T, but we also have to remove the energy of the flow work that was coming in.

CHAPTER 7
STUDY PROBLEMS

THE SECOND LAW OF THERMODYNAMICS

- **Heat engines and refrigerators**
- **The second law of thermodynamics**
- **The reversible and irreversible process**
- **The Carnot cycle and its efficiency**
- **The thermodynamic and ideal gas temperature scales**
- **Real machines and heat transfer with temperature differences**

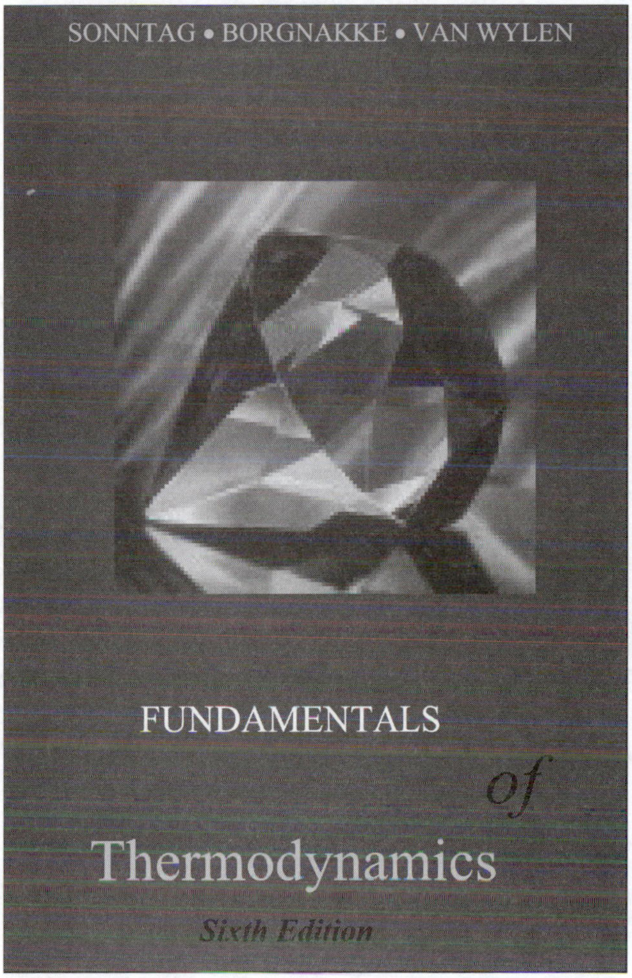

7.1 A window air-conditioner

A window air-conditioner is advertised as having a 5000 BTU/h cooling capacity. It requires an electrical power input of 500 W. Find the rate of heat rejected to the outside atmosphere and the coefficient of performance of the unit.

Solution:

C.V. Air-conditioner. Assume steady state so no storage of energy. The information provided is $\dot{W} = 500$ W and the heat absorbed is

$$\dot{Q}_L = 5000 \text{ BTU/h} = 5000 \times \frac{1.055}{3600} \text{ kW} = 1.4653 \text{ kW}$$

The energy equation gives:

$$\dot{Q}_H = \dot{Q}_L + \dot{W} = 1465.3 + 500 = \mathbf{1965\ W}$$

From the definition of the coefficient of performance, Eq.7.2

$$\beta_{REFRIG} = \frac{\dot{Q}_L}{\dot{W}} = \frac{1465.3}{500} = \mathbf{2.93}$$

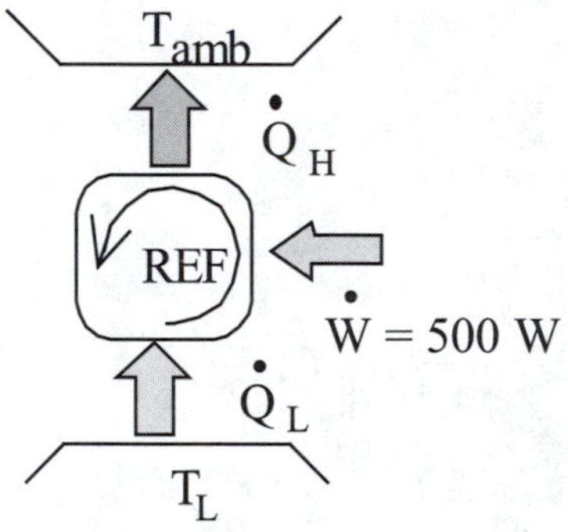

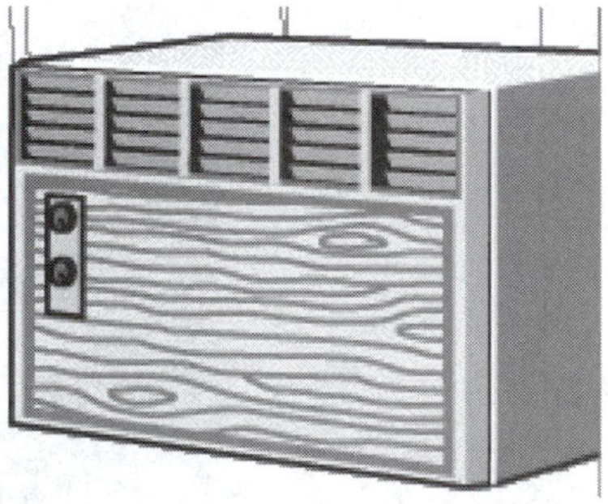

Remark: Assume the high and low temperatures in the cycle are 45°C and 5°C then a Carnot refrigerator would have

$$\beta_{REFRIG} = \frac{\dot{Q}_L}{\dot{W}} = \frac{T_L}{T_H - T_L} = \frac{278.15}{40} = \mathbf{6.95}$$

7.2 Influence of T_H on the heat engine efficiency

Let us examine a heat engine that rejects energy to the ambient, which is at $20^\circ C$. Assume it burns some fuel and air mixture providing heat addition temperatures of T_H = 1500 K, 1000 K and 750 K, respectively. What are the maximum expected thermal efficiencies?

Solution:

First, to consider the maximum efficiency we must assume the heat engine works as a Carnot heat engine. The efficiency then becomes, Eq.7.5

$$\eta_{Carnot\ HE} = \frac{W_{HE}}{Q_H} = 1 - \frac{T_L}{T_H}$$

from which we also see that the maximum is obtained with the minimum temperature for the rejection of energy T_L. For this we assume ambient so $T_L = T_{amb} = 20^\circ C = 293.15$ K and we can compute the three efficiencies

T_H :	750	1000	1500
$\eta_{Carnot\ HE}$	0.609	0.707	0.805

The efficiency increases with the high temperature approaching the upper limit of 1. Thus an amount of heat transfer at a higher temperature results in more useful work output than the same amount delivered at a lower temperature.

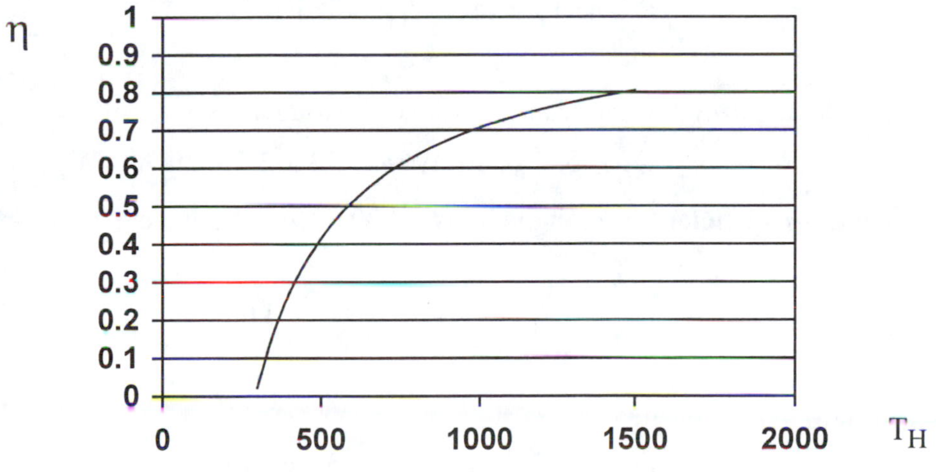

Comments: The low temperature was taken as the ambient T, which is unrealistic. In a car engine the exhaust gas temperature right out of the exhaust valve is higher than 750 K, similarly for a gas turbine or jet engine. With that low temperature in the Carnot cycle the efficiencies would be substantially lower, the lowest T_H would be efficiency of zero.

7.3 A domestic heat pump

A heat pump is used to heat a house during the winter. The motor that drives it is rated 2.5 kW. It is estimated that the house kept at 20°C will lose 14 kW when the outside temperature is – 5°C. Is the rated motor sufficient for an ideal heat pump? Suppose the actual heat pump has only one third the coefficient of performance as the ideal one. How much heating power could it supply?

Solution:

C.V. The house.

This is a control mass at steady state, $\dot{E}_{CV} = 0$.

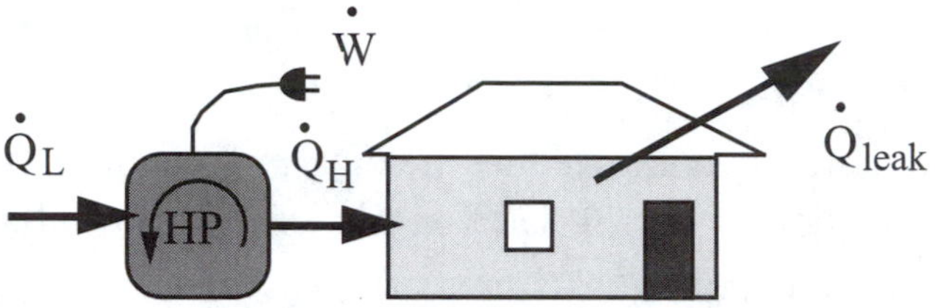

Energy Eq.5.31: $\dot{Q}_H = \dot{Q}_{leak}$

Definition of COP, Eq.7.16: $\beta_{HP} = \dfrac{\dot{Q}_H}{\dot{W}_{HP}} \le \beta_{Carnot\ HP}$

Second law Eq.7.13: $\beta_{Carnot\ HP} = \dfrac{T_H}{T_H - T_L} = \dfrac{20 + 273.15}{20 - (-5)} = 11.73$

So for the **ideal heat pump** we can receive a rate of heating as

$$\dot{Q}_H = \beta_{Carnot\ HP}\ \dot{W} = 11.73 \times 2.5 = \mathbf{29.33\ kW}$$

which is **more than sufficient** to keep the house at 20°C for which we need 14 kW.

Real heat pump: $\beta_{HP} = \dfrac{\dot{Q}_H}{\dot{W}_{HP}} = \beta_{Carnot\ HP}/3 = \dfrac{11.73}{3} = 3.91$

Now the heating rate becomes

$$\dot{Q}_H = \beta_{HP}\ \dot{W} = 3.91 \times 2.5 = \mathbf{9.78\ kW}$$

which is **not sufficient** to keep the house at 20°C.

7.4 An industrial heat pump for heat upgrade

In a large power plant 5 MW of 90°C waste-heat is the low temperature $\dot{Q}_L$ input to a heat pump that upgrades it to a $\dot{Q}_H$ delivered at 120°C. The process is said to occur with a power input of 0.4 MW. What does the second law say about this?

Solution:

C.V. The heat pump.

Energy Eq.: $\dot{Q}_H = \dot{Q}_L + \dot{W}_{HP}$

Definition of COP: $\beta_{HP} = \dfrac{\dot{Q}_H}{\dot{W}_{HP}} \le \beta_{\text{Carnot HP}}$

Second law: $\beta_{\text{Carnot HP}} = \dfrac{T_H}{T_H - T_L} = \dfrac{120 + 273.15}{120 - 90} = 13.1$

$$\dot{Q}_H = \dot{Q}_L + \dot{W} \;\Rightarrow\; \frac{\dot{Q}_L}{\dot{W}_{HP}} = \beta_{\text{Carnot HP}} - 1 = 13.1 - 1 = 12.1$$

$$\dot{W} = \dot{Q}_L / (\beta_{\text{Carnot HP}} - 1) = \frac{5000}{12.1} = \mathbf{413\ kW}$$

The stated 400 kW is thus **not sufficient**. Furthermore, we computed the power input requirement assuming an ideal heat pump. Any actual heat pump would have a lower COP and thus a higher required power input (probably 2-3 times as much)

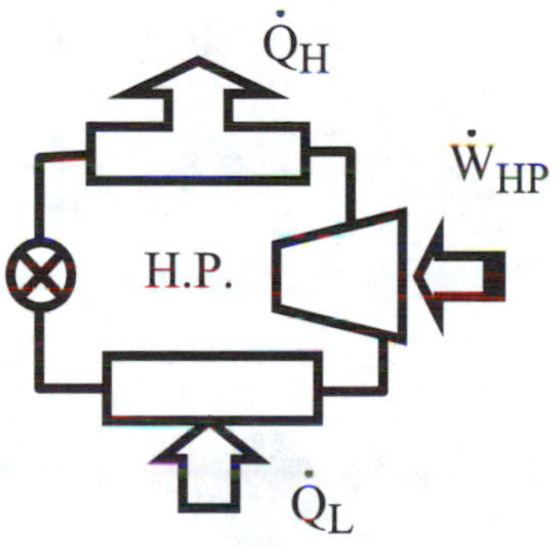

Comment: Instead of using any external work input to this process the waste heat can be divided into two with one fraction used in a heat engine rejecting energy to the ambient providing the work to drive the heat pump for the second part of the waste energy.

7.5 An air-conditioner unit

An air-conditioner should re-circulate 2 m^3/s air from the inside of a building at 25°C and deliver it back to the room at 15°C, thus providing cooling. What is the minimum amount of power input required? What mass flow rate of external air coming in at 30°C and exiting at 37°C is needed?

Solution:

In this application air is cooled to 15°C so we take $T_L = 15 + 273 = 288$ K and outside air is heated to $T_H = 37 + 273 = 310$ K.

C.V. Low temperature cooler section.

Energy equation Eq.6.12: $\dot{Q}_L = \dot{m}\,\Delta h \cong \dot{m}\,C_p\,\Delta T$

Volume flow rate given and assume ideal gas:

$$\dot{m}_{\text{air inside}} = \dot{V}/v = \frac{\dot{V}P}{RT} = \frac{2 \times 101.3}{0.287 \times 298} = 2.367 \text{ kg/s}$$

$$\dot{Q}_L = 2.367 \times 1.004 \times 10 = 23.765 \text{ kW}$$

COP for refrigerator Eq.7.15: $\beta_{\text{Carnot REF}} = \dfrac{T_L}{T_H - T_L} = \dfrac{288}{37 - 15} = 13.09$

$$\dot{W} = \frac{\dot{Q}_L}{\beta_{\text{Carnot REF}}} = \mathbf{1.82\ kW}$$

$$\dot{Q}_H = \dot{Q}_L + \dot{W} = 25.585 \text{ kW}$$

C.V. High temperature heater section.

Energy equation Eq.6.12: $\dot{Q}_H = \dot{m}\,\Delta h \cong \dot{m}\,C_p\,\Delta T$

$$\dot{m}_{\text{air outside}} = \frac{\dot{Q}_H}{C_p\,\Delta T} = \frac{25.585}{1.004\,(37 - 30)} = \mathbf{3.64\ kg/s}$$

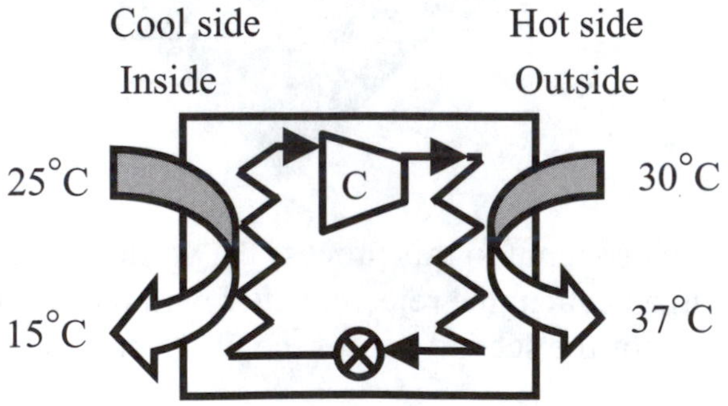

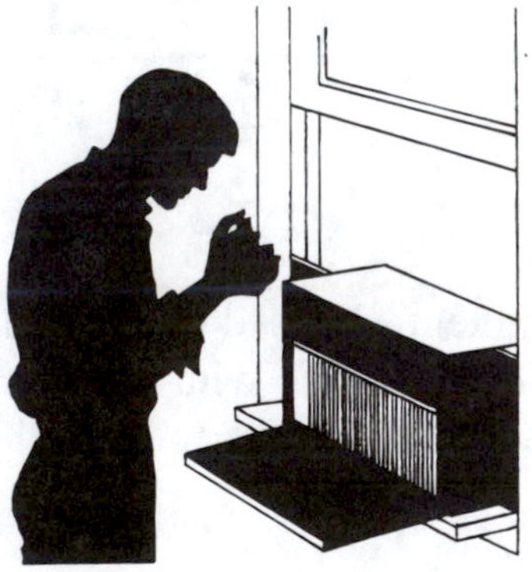

7.6 Making ice cubes

A tray for making ice-cubes is filled with 0.5 kg liquid water at 15°C. It is now put into the freezer where it is cooled down to – 10°C using 50 W of electrical power input. The freezer sits in a room at 25°C having a coefficient of performance as $\beta_{refrigerator} = \beta_{Carnot\ ref}/3$. How much energy is used as work input to accomplish this process? If we neglect other cooling loads how much time will this process take?

Solution:

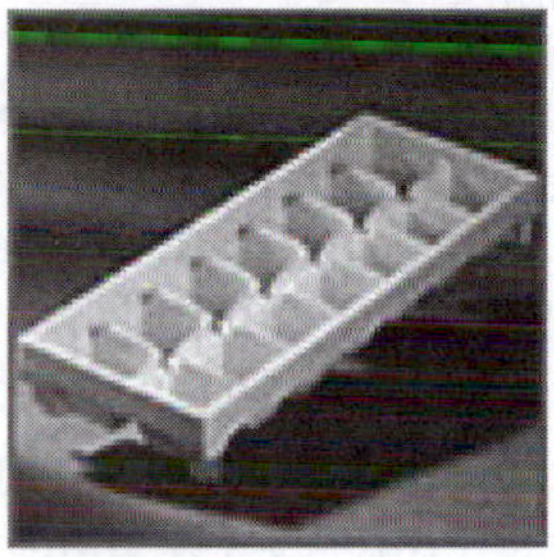

C.V. Water in the tray.
This is a control mass. As the water freezes it happens at constant pressure of 101 kPa.

Energy Eq.: $m(u_2 - u_1) = {}_1Q_2 - {}_1W_2$

State 1: Compressed liquid, use saturated liquid same T, from Table B.1.1

$u_1 = 62.98$ kJ/kg, $v_1 = 0.001001$ m^3/kg

State 2: Compressed solid, use saturated solid same T, from Table B.1.5

$u_2 = -354.09$ kJ/kg, $v_2 = 0.0010891$ m^3/kg

Process: $P = C$; ${}_1W_2 = P\, m(v_2 - v_1) = 101 \times 0.5(0.0010891 - 0.001001) = 0.0044$ kJ

From the energy equation

$${}_1Q_2 = m(u_2 - u_1) + {}_1W_2 = 0.5\,(-354.09 - 62.98) + 0.0044 = -208.5 \text{ kJ}$$

Notice how small the work is; it could have been neglected. The heat transfer out of the water is then into the cold air in the freezer which is the Q_L into the refrigerator cycle.

The Carnot cycle refrigerator has a coefficient of performance COP as, Eq.7.15

$$\beta_{Carnot\ ref} = \frac{T_L}{T_H - T_L} = \frac{273 - 10}{25 - (-10)} = 7.51$$

and the actual refrigerator has

$$\beta_{refrigerator} = \frac{\beta_{Carnot\ ref}}{3} = 2.5$$

Heat transfer out of the water is into the refrigerator cycle, so

$$W = \frac{-{}_1Q_2}{\beta_{refrigerator}} = \frac{208.5}{2.5} = \mathbf{83.4\ kJ}$$

With the constant rate we have $W = \Delta t\, \dot{W}$ so the elapsed time is

$$\Delta t = \frac{W}{\dot{W}} = \frac{83.4 \times 1000}{50}\ \frac{\text{kJ} \times \text{J/kJ}}{\text{J/s}} = \mathbf{1668\ s = 27.8\ min}$$

7.7 Finite temperature difference heat transfer

A meat warehouse kept at $T_L = -18^\circ C$ by a commercial refrigeration system gains energy by heat transfer from the outside at a rate $\dot{Q} = B\,(T_H - T_L)$. Assume B = 1.1 kW/K determined by the wall construction, insulation etc. The actual refrigerator has a coefficient of performance that equals ¼ of the corresponding Carnot refrigerator and it has a 9.5 kW motor to run the system. How high an outside temperature can be allowed?

Solution:

The higher the outside temperature the higher the heat transfer leak into the cold space is and the coefficient of performance drops due to the higher temperature difference. There is thus a limit for the outside temperature for which the 9.5 kW motor is sufficient.

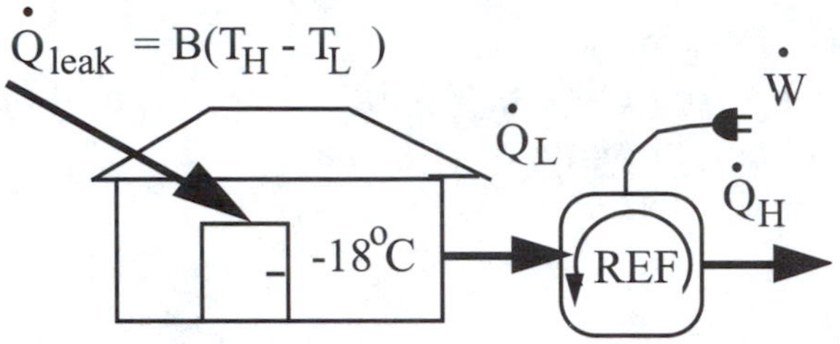

The Carnot cycle refrigerator has a coefficient of performance COP as, Eq.7.15 and the actual one is ¼ thereof.

$$\beta_{\text{Carnot ref}} = \frac{T_L}{T_H - T_L} \; ; \qquad \beta_{\text{refrigerator}} = \frac{\beta_{\text{Carnot ref}}}{4} = \frac{\dot{Q}_L}{\dot{W}}$$

At steady state we must remove the heat transfer leak from the cold space so

$$\dot{Q}_L = \dot{Q}_{\text{leak}} = B\,(T_H - T_L)$$

The cycle with the given power input can provide

$$\dot{Q}_L = \beta_{\text{refrigerator}}\,\dot{W}$$

Setting these equal and substitute in the coefficient of performance we get

$$B\,(T_H - T_L) = \beta_{\text{refrigerator}}\,\dot{W} = \frac{\beta_{\text{Carnot ref}}}{4}\,\dot{W} = \frac{T_L}{T_H - T_L}\,\frac{1}{4}\,\dot{W}$$

and now rearrange the equation to get

$$(T_H - T_L)^2 = \frac{T_L}{4\,B}\,\dot{W} = \frac{273 - 18}{4 \times 1.1}\,\frac{\text{K}}{\text{kW/K}} \times 9.5\ \text{kW} = 550.6\ \text{K}^2$$

solve

$$T_H - T_L = \sqrt{550.6} = 23.5\ \text{K} \quad \Rightarrow \quad T_H = 23.5 - 18 = \mathbf{5.5^\circ C}$$

This of course is not very warm and the refrigerator is not sized large enough.

7.3E A domestic heat pump

A heat pump is used to heat a house during the winter. The motor that drives it is rated 2.5 kW. It is estimated that the house kept at 68 F will lose 14 Btu/s when the outside temperature is 22 F. Is the rated motor sufficient for an ideal heat pump? Suppose the actual heat pump has only one third the coefficient of performance as the ideal one. How much heating power could it supply?

Solution:

C.V. The house.

This is a control mass at steady state, $\dot{E}_{CV} = 0$.

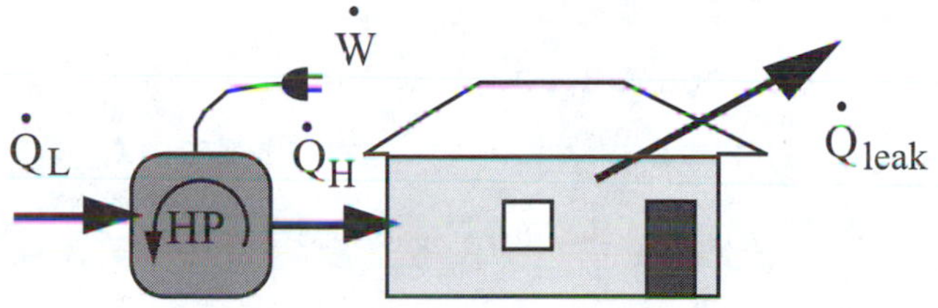

Energy Eq.5.31: $\dot{Q}_H = \dot{Q}_{leak}$

Definition of COP, Eq.7.16: $\beta_{HP} = \dfrac{\dot{Q}_H}{\dot{W}_{HP}} \leq \beta_{Carnot\ HP}$

Second law Eq.7.13: $\beta_{Carnot\ HP} = \dfrac{T_H}{T_H - T_L} = \dfrac{68 + 459.67}{68 - 22} = 11.47$

So for the **ideal heat pump** we can receive a rate of heating as

$$\dot{Q}_H = \beta_{Carnot\ HP}\ \dot{W} = 11.47 \times 2.5 = \mathbf{28.675\ kW = 27.18\ Btu/s}$$

which is **more than sufficient** to keep the house at 68 F for which we need 14 Btu/s.

Real heat pump: $\beta_{HP} = \dfrac{\dot{Q}_H}{\dot{W}_{HP}} = \beta_{Carnot\ HP}/3 = \dfrac{11.47}{3} = 3.82$

Now the heating rate becomes

$$\dot{Q}_H = \beta_{HP}\ \dot{W} = 3.82 \times 2.5 = \mathbf{9.55\ kW = 9.05\ Btu/s}$$

which is **not sufficient** to keep the house at 68 F.

7.5E An air-conditioner unit

An air-conditioner should re-circulate 70 ft^3/s air from the inside of a building at 77 F and deliver it back to the room at 59 F, thus providing cooling. What is the minimum amount of power input required? What mass flow rate of external air coming in at 86 F and exiting at 100 F is needed?

Solution:

In this application air is cooled to 59 F so we take $T_L = 59 + 459.67 = 518.7$ R and outside air is heated to $T_H = 100 + 459.67 = 559.7$ R.

C.V. Low temperature cooler section.

Energy equation Eq.6.12: $\dot{Q}_L = \dot{m}\,\Delta h \cong \dot{m}\,C_p\,\Delta T$

Volume flow rate given and assume ideal gas:

$$\dot{m}_{air\ inside} = \dot{V}/v = \frac{\dot{V}P}{RT} = \frac{70 \times 14.7 \times 144}{53.34 \times 536.67} = 5.176 \text{ lbm/s}$$

$$\dot{Q}_L = 5.176 \times 0.24 \times (77 - 59) = 22.36 \text{ Btu/s}$$

COP for refrigerator Eq.7.15: $\beta_{Carnot\ REF} = \dfrac{T_L}{T_H - T_L} = \dfrac{518.7}{100 - 59} = 12.65$

$$\dot{W} = \frac{\dot{Q}_L}{\beta_{Carnot\ REF}} = \mathbf{1.77\ Btu/s}$$

$$\dot{Q}_H = \dot{Q}_L + \dot{W} = 24.13 \text{ Btu/s}$$

C.V. High temperature heater section.

Energy equation Eq.6.12: $\dot{Q}_H = \dot{m}\,\Delta h \cong \dot{m}\,C_p\,\Delta T$

$$\dot{m}_{air\ outside} = \frac{\dot{Q}_H}{C_p\,\Delta T} = \frac{24.13}{0.24\,(100 - 86)} = \mathbf{7.18\ lbm/s}$$

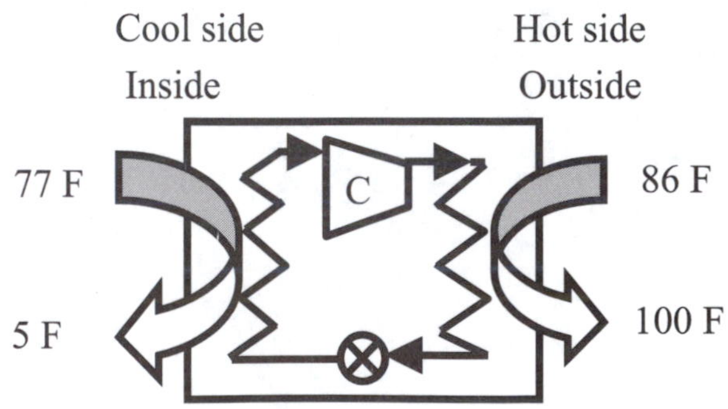

7.6E Making ice cubes

A tray for making ice-cubes is filled with 1 lbm liquid water at 60 F. It is now put into the freezer where it is cooled down to 15 F using 50 W of electrical power input. The freezer sits in a room at 77 F having a coefficient of performance as $\beta_{refrigerator} = \beta_{Carnot\ ref}/3$. How much energy is used as work input to accomplish this process? If we neglect other cooling loads how much time will this process take?

Solution:

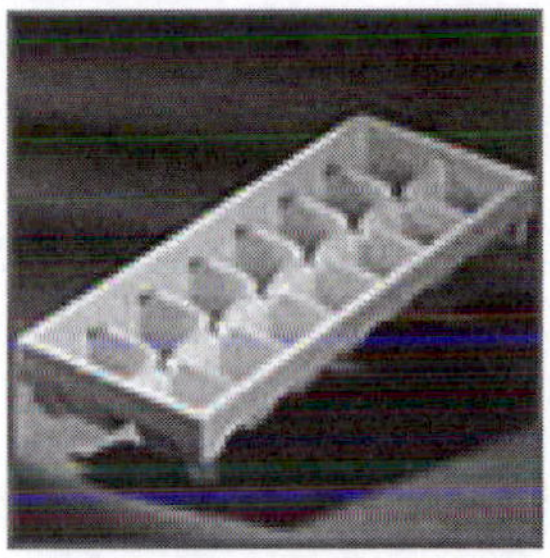

C.V. Water in the tray.
This is a control mass. As the water freezes it happens at constant pressure of 14.7 psia.

Energy Eq.: $m(u_2 - u_1) = {}_1Q_2 - {}_1W_2$

State 1: Compressed liquid, use saturated liquid same T, from Table F.7.1

$u_1 = 28.08$ Btu/lbm, $v_1 = 0.01603$ ft^3/lbm

State 2: Compressed solid, use saturated solid same T, from Table F.7.4

$u_2 = -151.75$ Btu/lbm, $v_2 = 0.01745$ ft^3/lbm

Process: P = C; ${}_1W_2 = P\,m(v_2 - v_1) = 14.7 \times 144 \times 1\,(0.01745 - 0.01603)$

$= 3.006$ lbf-ft $= 0.00386$ Btu

From the energy equation

$${}_1Q_2 = m(u_2 - u_1) + {}_1W_2 = 1\,(-151.75 - 28.08) + 0.00386 = -179.8 \text{ Btu}$$

Notice how small the work is; it could have been neglected. The heat transfer out of the water is then into the cold air in the freezer which is the Q_L into the refrigerator cycle.

The Carnot cycle refrigerator has a coefficient of performance COP as, Eq.7.15

$$\beta_{Carnot\ ref} = \frac{T_L}{T_H - T_L} = \frac{459.67 + 15}{77 - 15} = 7.656$$

and the actual refrigerator has

$$\beta_{refrigerator} = \frac{\beta_{Carnot\ ref}}{3} = 2.55$$

Heat transfer out of the water is into the refrigerator cycle, so

$$W = \frac{-{}_1Q_2}{\beta_{refrigerator}} = \frac{179.8}{2.55} = \mathbf{70.5\ Btu}$$

With the constant rate we have $W = \Delta t\,\dot{W}$ so the elapsed time is

$$\Delta t = \frac{W}{\dot{W}} = \frac{70.5 \times 3600}{50 \times 3.412}\,\frac{\text{Btu} \times \text{s/h}}{\text{Btu/h}} = \mathbf{1488\ s = 24.8\ min}$$

CHAPTER 8
STUDY PROBLEMS

ENTROPY

- **The inequality of Clausius**
- **The property entropy**
- **Entropy changes in processes**
- **Entropy generation**
- **Entropy changes for solids and liquids**
- **Entropy changes for an ideal gas**
- **The reversible polytropic process**

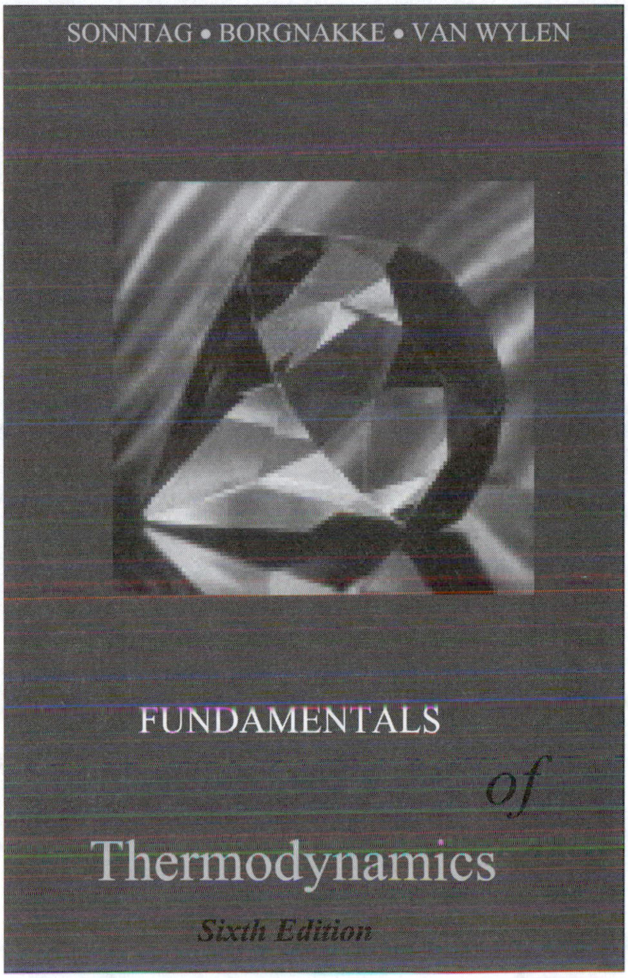

8.1 A heat engine efficiency from the inequality of Clausius

Consider an actual heat engine with efficiency of η working between reservoirs at T_H and T_L. Use the second law in the form of Eq.8.1 for the whole heat engine to prove that

a. If the heat engine is ideal then $\eta = 1 - \frac{T_L}{T_H}$

b. If the heat engine is non-ideal then $\eta \le \eta_{\text{Carnot HE}}$

Solution:

C.V. Heat engine out to the reservoirs at T_H and T_L. The heat engine whether ideal or not does not store any energy or entropy so the energy equation is

Energy Eq.: $0 = Q_H - Q_L - W$

Clausius : $\oint \frac{dQ}{T} = \frac{Q_H}{T_H} - \frac{Q_L}{T_L} \le 0 \quad \Rightarrow \quad Q_L \ge Q_H \frac{T_L}{T_H}$

Substitute this into the energy equation and solve for the work

$$W = Q_H - Q_L \le Q_H \left[1 - \frac{T_L}{T_H} \right]$$

Case a. Ideal, the equal sign applies and the work is

$$W = Q_H \left[1 - \frac{T_L}{T_H} \right] = Q_H \eta$$

and the efficiency is equal to the Carnot cycle efficiency.

Case b. Non-ideal, the inequality applies and the efficiency $\eta = W/Q_H$ becomes

$$\eta \le \left[1 - \frac{T_L}{T_H} \right] = \eta_{\text{Carnot HE}}$$

And we see that the efficiency is smaller than the Carnot efficiency.

Remark: After writing the entropy equation as Eq.8.11 we can express how much smaller the efficiency is due to S_{gen} as

Entropy Eq.8.11: $0 = \oint \frac{dQ}{T} + \oint \delta S_{gen} = \frac{Q_H}{T_H} - \frac{Q_L}{T_L} + S_{gen} \; ; \; S_{gen} \ge 0$

$$0 = \frac{Q_H}{T_H} - \frac{Q_L}{T_L} + S_{gen} \quad \Rightarrow \quad Q_L = Q_H \frac{T_L}{T_H} + T_L S_{gen}$$

Notice that Q_L is larger than the ideal case by an amount proportional to S_{gen}

$$W = Q_H - Q_L - T_L S_{gen} = Q_H \left[1 - \frac{T_L}{T_H} \right] - T_L S_{gen} = Q_H \eta$$

$$\eta = \left[1 - \frac{T_L}{T_H} \right] - \frac{T_L S_{gen}}{Q_H}$$

8.2 Change in s from the steam tables

Water at 100°C is heated to 150°C at constant P. What are changes in u and s when the starting state is at

a. 10 000 kPa b. 500 kPa c. 100 kPa d. 10 kPa

Solution:

a. From Table B.1.4 compressed liquid states at 10 000 kPa, we get

$$u_2 - u_1 = 627.39 - 416.09 = \mathbf{211.3\ kJ/kg}$$

$$s_2 - s_1 = 1.8304 - 1.2992 = \mathbf{0.5312\ kJ/kg\ K}$$

b. From Table B.1.4 compressed liquid states at 500 kPa, we get

$$u_2 - u_1 = 631.66 - 418.8 = \mathbf{212.9\ kJ/kg}$$

$$s_2 - s_1 = 1.8415 - 1.3065 = \mathbf{0.535\ kJ/kg\ K}$$

c. From Table B.1.3 superheated vapor states at 100 kPa, we get

$$u_2 - u_1 = (2582.75 - 2506.06)\frac{50}{150 - 99.62} = \mathbf{76.11\ kJ/kg}$$

$$s_2 - s_1 = (7.6133 - 7.3593)\frac{50}{150 - 99.62} = \mathbf{0.252\ kJ/kg\ K}$$

d. From Table B.1.3 superheated vapor states at 10 kPa, we get

$$u_2 - u_1 = 2587.86 - 2515.5 = \mathbf{72.1\ kJ/kg}$$

$$s_2 - s_1 = 8.6881 - 8.4479 = \mathbf{0.2402\ kJ/kg\ K}$$

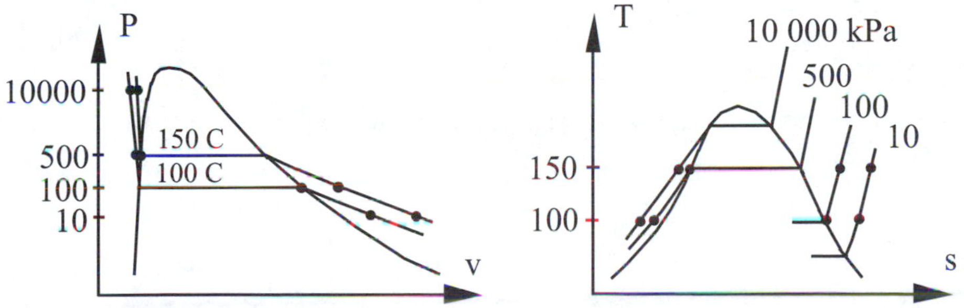

The changes in u and s are larger for the liquid than for the gas. Within the liquid phase the influence of P is very modest. Water vapor at 100 kPa is close to an ideal gas ($Z = Pv_g/RT_{sat} = 0.985$). In the vapor phase changing P (fixed T) from 100 kPa to 10 kPa changes s, but u is nearly constant. The influence of P becomes much stronger as the state approaches the dense fluid region near the critical point.

8.3 Direction of work and heat transfer

Ammonia at 200 kPa, x = 1.0 is compressed in a piston/cylinder to 1 MPa, 100°C in a reversible process. Find the sign for the work and the sign for the heat transfer.

Solution:

The directions of work and heat transfer are given by variations in v and s as:

Work Eq.4.3: $_1w_2 = \int P\, dv$ so sign $_1w_2$ follows sign of dv

Heat transfer Eq.8.2: $_1q_2 = \int T\, ds$ so sign $_1q_2$ follows sign of ds

Properties from the ammonia tables:

State 1: Table B.2.2: $v_1 = 0.5946\ m^3/kg$; $s_1 = 5.5979$ kJ/kg K

State 2: Table B.2.2: $v_2 = 0.1739\ m^3/kg$; $s_2 = 5.6342$ kJ/kg K

Changes in v and s gives:

$dv < 0$ $\Rightarrow$ **w is negative**

$ds > 0$ $\Rightarrow$ **q is positive**

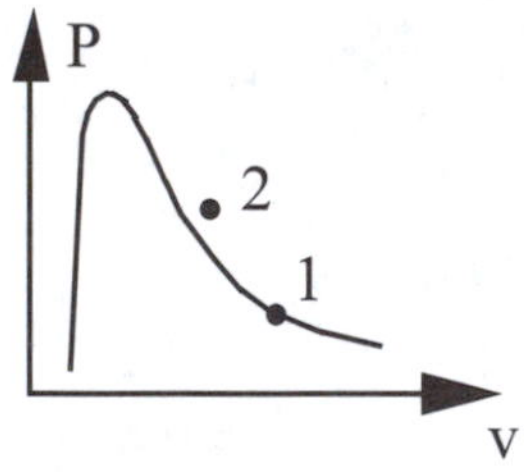

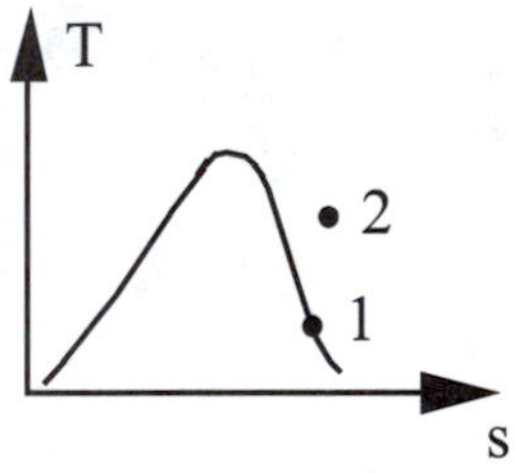

Remark: Since the process (path) was not specified we do not know the amounts of $_1w_2$ and $_1q_2$

8.4 An internally reversible, externally irreversible process

Ammonia in a piston-cylinder that maintains constant P is at 20°C, 1600 kPa and is now heated to 60°C in an internally reversible process. The external heat source is at a constant 70°C. We want to find the specific heat transfer in the process and the external (total) specific entropy generation.

Solution:

Take as a control volume the ammonia only, this is a control mass. Neglect storage of energy and entropy in the cylinder walls and piston mass.

Energy Eq.: $u_2 - u_1 = {}_1q_2 - {}_1w_2$

Entropy Eq.: $s_2 - s_1 = \int_1^2 \frac{dq}{T} + {}_1s_{2\ gen\ ammonia}$

Process: $P = C \Rightarrow {}_1w_2 = \int P\, dv = P\,(v_2 - v_1)$

Reversible $\Rightarrow {}_1s_{2\ gen\ ammonia} = 0$

State 1: Table B.2.1 Approximate compressed liquid state with saturated same T

$$v_1 = 0.001638\ m^3/kg, \quad u_1 = 272.89\ kJ/kg, \quad s_1 = 1.0408\ kJ/kg\ K$$

State 2: Table B.2.2 Superheated vapor

$$v_2 = 0.08951\ m^3/kg, \quad u_2 = 1389.3\ kJ/kg, \quad s_2 = 5.0472\ kJ/kg\ K$$

Work from the process equation is

$${}_1w_2 = P\,(v_2 - v_1) = 1600\ kPa\ (0.08951 - 0.001638)\ m^3/kg = 140.6\ kJ/kg$$

Heat transfer is from the energy equation (we could also have done it as $h_2 - h_1$)

$${}_1q_2 = u_2 - u_1 + {}_1w_2 = 1389.3 - 272.89 + 140.6 = \mathbf{1257\ kJ/kg}$$

The external entropy generation is a q over a ΔT so we can do it as in Eq.8.18:

$${}_1s_{2\ gen\ tot} = \Delta s_{CV} + \Delta s_{SUR} = s_2 - s_1 + [-\frac{{}_1q_2}{T_{source}}]$$

$$= 5.0472 - 1.0408 - \frac{1257}{273.15 + 70} = \mathbf{0.343\ kJ/kg\ K}$$

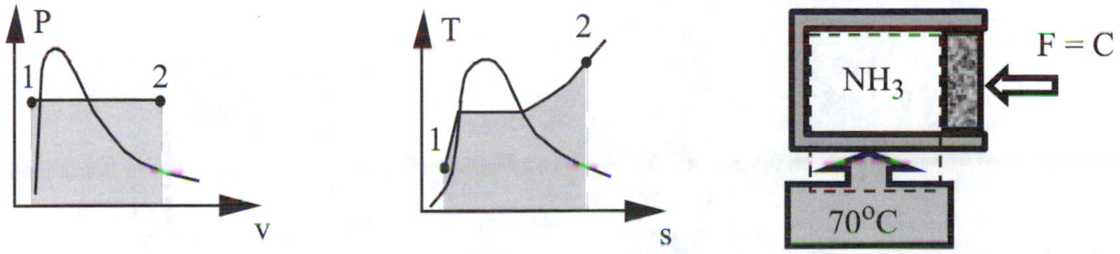

Choosing a CV that is the ammonia plus walls out to the source has the heat transfer crossing the CS at T_{source} not the ammonia T so the entropy equation becomes

$$s_2 - s_1 = \int_1^2 \frac{dq}{T_s} + {}_1s_{2\ gen\ tot} = \frac{{}_1q_2}{T_{source}} + {}_1s_{2\ gen\ tot}$$

and as the space where T changes (s is generated) is included in the CV, the entropy generation term includes that too becoming the same as before.

8.5 An irreversible mixing process

A piston cylinder with a constant force on it contains 0.25 kg R-134a saturated vapor in volume A and 0.1 kg of R-134a at 60°C in volume B both at 600 kPa. The two masses mix in an adiabatic process. We want to know the final temperature and the entropy generation for the process.

Solution:

Continuity Eq.: $m_2 - m_A - m_B = 0$

Energy Eq.5.11: $m_2u_2 - m_Au_A - m_Bu_B = -{}_1W_2$

Entropy Eq.8.14: $m_2s_2 - m_As_A - m_Bs_B = \int dQ/T + {}_1S_{2\ gen}$

Process: $P = \text{Constant} \;\; => \;\; {}_1W_2 = \int PdV = P(V_2 - V_1)$

$Q = 0$

Substitute the work term into the energy equation and rearrange to get

$$m_2u_2 + P_2V_2 = m_2h_2 = m_Au_A + m_Bu_B + PV_1 = m_Ah_A + m_Bh_B$$

where the last rewrite used $PV_1 = PV_A + PV_B$.

State A1: Table B.5.2 $h_A = 410.66$ kJ/kg ; $s_A = 1.7179$ kJ/kg K

State B1: Table B.5.2 $h_B = 448.28$ kJ/kg ; $s_B = 1.8379$ kJ/kg K

Energy equation gives:

$$h_2 = \frac{m_A}{m_2}h_A + \frac{m_B}{m_2}h_B = \frac{0.25}{0.35}410.66 + \frac{0.1}{0.35}448.28 = 421.41 \text{ kJ/kg}$$

State 2: $(P_2, h_2) \;\; \Rightarrow \;\; s_2 = 1.7536$ kJ/kg K; $T_2 = \mathbf{32.4°C}$

With the zero heat transfer we have

$${}_1S_{2\ gen} = m_2s_2 - m_As_A - m_Bs_B$$
$$= 0.35 \times 1.7536 - 0.25 \times 1.7179 - 0.10 \times 1.8379 = \mathbf{0.0005\ kJ/K}$$

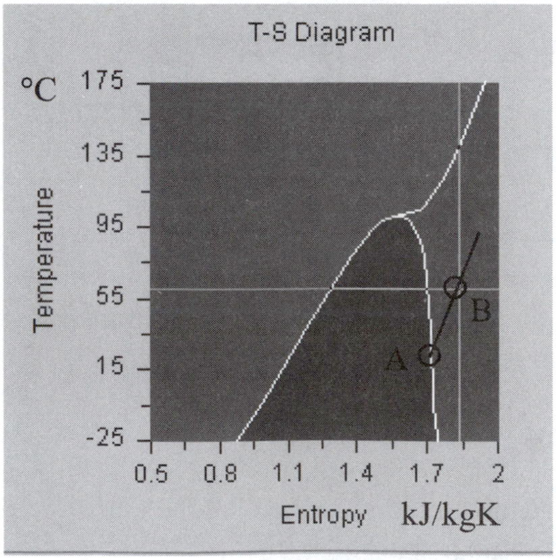

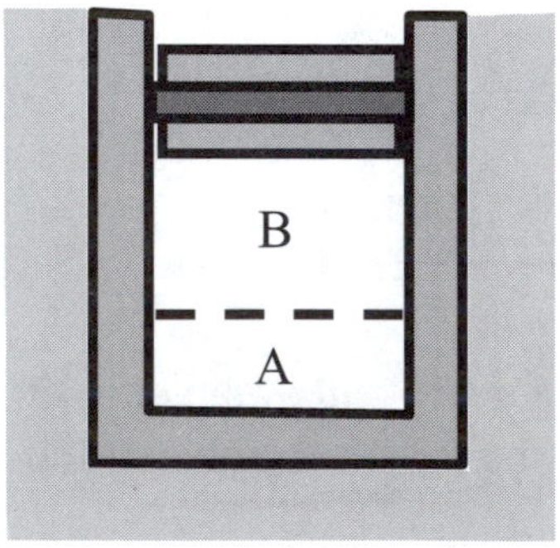

8.6 Cooling of a solid with water

Due to a sensor malfunction a 100 kg steel tank has been heated to 250°C. To cool it down we put **x** kg of liquid water at 20°C into it and wait for the masses to come to final uniform temperature of maximum 80°C. Assume the tank is open so P = 100 kPa and neglect any external heat transfer. Find the required water mass **x** and the total entropy generation in the process assuming no water evaporates.

Solution:

C.V. Steel tank and contents, constant pressure process

Energy Eq.: $m_{steel}(u_2 - u_1)_{steel} + m_{H_2O}(u_2 - u_1)_{H_2O} = {}_1Q_2 - {}_1W_2$

Entropy Eq.: $m_{steel}(s_2 - s_1)_{steel} + m_{H_2O}(s_2 - s_1)_{H_2O} = \int dQ/T + {}_1S_{2\ gen}$

Process Eq.: $P = \text{constant}$ and ${}_1Q_2 = 0$

$$\Rightarrow \quad {}_1W_2 = P(V_2 - V_1)$$

Substitute the work term into the energy equation and get:

$$\Rightarrow \quad m_{steel}(h_2 - h_1)_{steel} + m_{H_2O}(h_2 - h_1)_{H_2O} = 0$$

For this problem we may also say that the work is nearly zero as the steel and the liquid water will not change volume to any measurable extent. Now we get changes in u's instead of h's. For these phases we have $C_V = C_P = C$ from Tables A.3 and A.4.

Table A.3: $C_{steel} = 0.46$ kJ/kg K   Table A.4: $C_{H_2O} = 4.18$ kJ/kg K

The energy equation with $m_{H_2O} = \mathbf{x}$ becomes

$$100 \times 0.46 \times (80 - 250) + \mathbf{x} \times 4.18 \times (80 - 20) = 0$$

$$\mathbf{x = 31.18\ kg}$$

The entropy generation from the entropy equation is, using Eq.8.20,

$${}_1S_{2\ gen} = m_{steel}(s_2 - s_1)_{steel} + m_{H_2O}(s_2 - s_1)_{H_2O}$$

$$= 100 \times 0.46 \ln\left(\frac{353.15}{523.15}\right) + 31.18 \times 4.18 \ln\left(\frac{353.15}{293.15}\right) = \mathbf{6.19\ kJ/K}$$

Conceptual schematics:

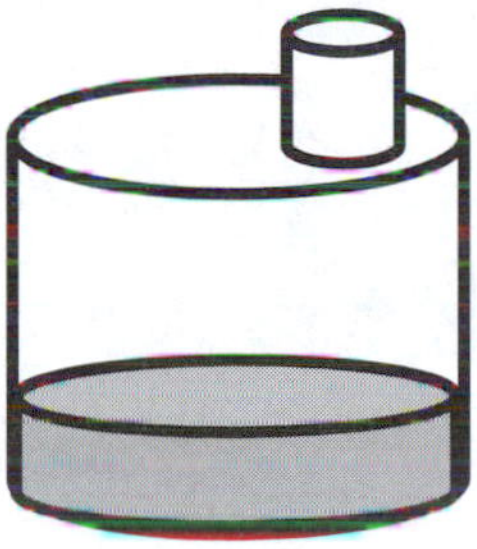

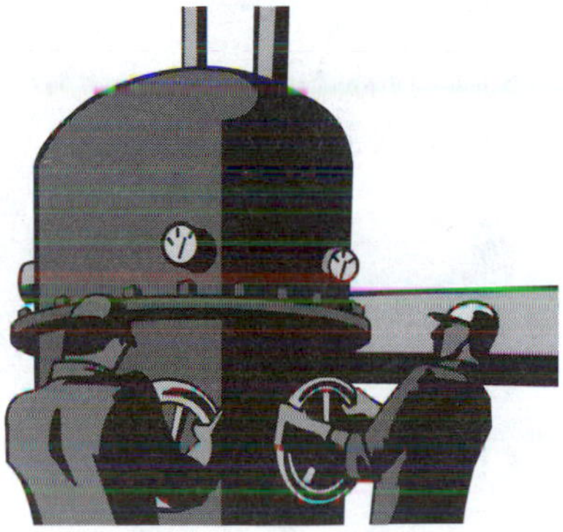

8.7 An isothermal expansion of air

A mass of 1 kg of air contained in a cylinder at 1.5 MPa, 1000 K, expands in a reversible isothermal process to a pressure 5 times smaller. Calculate the change of entropy of the air and the heat transfer during the process.

Solution:

C.V. Air, which is a control mass.

Energy Eq.5.11: $m(u_2 - u_1) = {}_1Q_2 - {}_1W_2$

Entropy Eq.8.14: $m(s_2 - s_1) = \int dQ/T + {}_1S_{2\ gen}$

Process: $T = \text{constant}$ so $\int dQ/T = {}_1Q_2 /T$

reversible implies ${}_1S_{2\ gen} = 0$

State 1: P_1, T_1 State 2: $T_2 = T_1,\ v_2 = v_1/5$

The change of entropy for the air (ideal gas) is from Eq.8.26

$$\Delta S_{air} = m(s_2 - s_1) = m[C_{vo} \ln \frac{T_2}{T_1} + R \ln \frac{v_2}{v_1}] = 1\ [\ 0 + 0.287 \ln(5)] = \mathbf{0.462\ kJ/K}$$

So the heat transfer from the second law becomes

$${}_1Q_2 = T\ m\ (s_2 - s_1) = 1000\ K \times 0.462\ kJ/K = \mathbf{462\ kJ}$$

Comment: For a non-ideal gas we would find state 2 in the tables and get (u_2, s_2), first do ${}_1Q_2$ and then do ${}_1W_2$ from the energy equation

$${}_1W_2 = {}_1Q_2 - m(u_2 - u_1)$$

For an ideal gas constant T => $u_2 = u_1$ then ${}_1W_2 = {}_1Q_2$

From the process equation and ideal gas law

$$PV = mRT = \text{constant}$$

we can calculate the work term (polytropic process) as in Eq.4.5

$${}_1Q_2 = {}_1W_2 = \int PdV = P_1V_1 \ln (V_2/V_1) = mRT_1 \ln (V_2/V_1)$$

$$= 1 \times 0.287 \times 1000 \ln (5) = 461.91\ kJ$$

consistent with the above result.

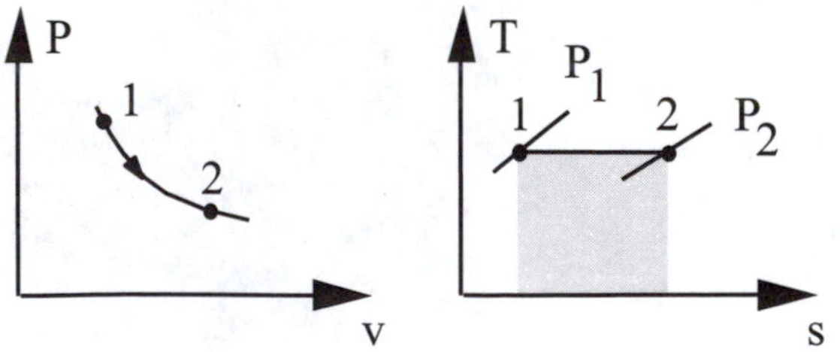

8.8 A light bulb filled with argon gas

A light bulb has been on a while so 0.5 g of Argon and 50 g of glass (we disregard the rest of bulb) is at 80°C. The bulb is turned off and now cools to ambient temperature of 15°C. The Argon was at 110 kPa while warm. Find the total entropy generation in the process.

Solution:

C.V. Argon gas and the glass out to the ambient. Constant volume so W = 0.

Energy Eq.5.11: $U_2 - U_1 = m_{argon}(u_2 - u_1) + m_{glass}(u_2 - u_1) = {}_1Q_2$

Entropy Eq.8.14: $S_2 - S_1 = \int dQ/T + {}_1S_{2\,gen} = {}_1S_{2\,gen} + {}_1Q_2/T_2$

Process: v = constant => For argon gas: $P_2/P_1 = T_2/T_1$

The heat transfer becomes the change in energy of the argon and the glass

$$
\begin{aligned}
{}_1Q_2 &= m_{argon}(u_2 - u_1) + m_{glass}(u_2 - u_1) \\
&= m_{argon}C_{argon}(T_2 - T_1) + m_{glass}C_{glass}(T_2 - T_1) \\
&= 0.0005 \times 0.312 \times (-65) + 0.050 \times 0.8 \times (-65) \\
&= -0.0101 - 2.6 = -2.61 \text{ kJ}
\end{aligned}
$$

Evaluate changes in s from Eq.8.25 or 8.26 for the argon gas

$$
\begin{aligned}
m_{argon}(s_2 - s_1) &= mC_p \ln(T_2/T_1) - mR\ln(T_2/T_1) = mC_v \ln(T_2/T_1) \\
&= 0.0005 \times 0.312 \ln\left[\frac{288.15}{353.15}\right] = -0.0000317 \text{ kJ/K}
\end{aligned}
$$

Evaluate the change in s for the solid glass, Eq.8.20

$$
m_{glass}(s_2 - s_1) = mC\ln\frac{T_2}{T_1} = 0.05 \times 0.8 \ln\left[\frac{288.15}{353.15}\right] = -0.00814 \text{ kJ/K}
$$

Now the entropy generation becomes

$$
\begin{aligned}
{}_1S_{2\,gen} &= S_2 - S_1 - {}_1Q_2/T_2 = m_{argon}(s_2 - s_1) + m_{glass}(s_2 - s_1) - {}_1Q_2/T_2 \\
&= -0.0000317 - 0.00814 - (-2.61/288.15) = \mathbf{0.00089\ kJ/K}
\end{aligned}
$$

The heat transfer from the argon goes out through the glass and the total heat transfer from both masses goes out through a convection layer in the air. The entropy generation is distributed but our lumped (overall) analysis cannot give this information.

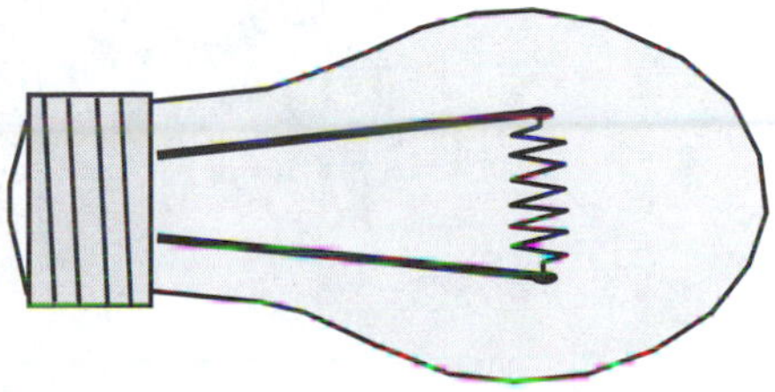

8.9 Entropy generation and its location due to heat transfer

Hot combustion gases at 1500 K loses 3000 W to the steel walls at 700 K which in turn delivers the 3000 W to a flow of glycol at 100°C and the glycol has heat transfer to atmospheric air at 20°C. Assume steady state and find the rates of entropy generation and where it is generated.

Solution:

For a C.V at steady state we have the entropy equation as a rate form as Eq.8.43

$$\frac{dS_{c.v.}}{dt} = 0 = \int d\dot{Q}/T + \dot{S}_{gen}$$

CV1. From combustion gases at 1500 K to the steel at 700 K. $\dot{Q}$ goes through but enters and leaves at two different T's

$$\dot{S}_{gen1} = -\int d\dot{Q}/T = \frac{-3000}{1500} - [\frac{-3000}{700}] = \mathbf{2.286\ W/K}$$

CV2. From the steel wall at 700 K to the glycol at 100°C. $\dot{Q}$ goes through but enters and leaves at two different T's

$$\dot{S}_{gen2} = -\int d\dot{Q}/T = [\frac{-3000}{700}] - [\frac{-3000}{373.15}] = \mathbf{3.754\ W/K}$$

CV3 (includes the glycol flow). From the glycol at 100°C to the atmospheric air at 20°C. $\dot{Q}$ goes through but enters and leaves at two different T's

$$\dot{S}_{gen3} = -\int d\dot{Q}/T = [\frac{-3000}{373.15}] - [\frac{-3000}{293.15}] = \mathbf{2.194\ W/K}$$

Notice the biggest $\dot{S}_{gen}$ is for the largest change $\Delta[1/T]$

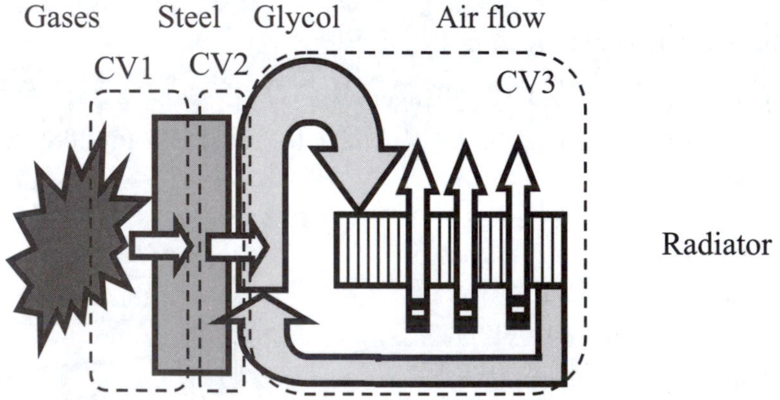

Remark: The flux of S is $\dot{Q}/T$ flowing across a surface. Notice how this flux increases as $\dot{Q}$ flows towards lower and lower T.

T [K]	1500	700	373.15	293.15
$\dot{Q}/T$ [kW/K]	2	4.286	8.04	10.23

8.2E Change in s from the steam tables

Water at 200 F is heated to 400 F at constant P. What are changes in u and s when the starting state is at

a. 2000 psia b. 500 psia c. 10 psia d. 1 psia

Solution:

a. From Table F.7.3 compressed liquid states at 2000 psia, we get

$u_2 - u_1 = 370.38 - 166.48 = \mathbf{203.9\ Btu/lbm}$

$s_2 - s_1 = 0.5621 - 0.2916 = \mathbf{0.2705\ Btu/lbm\ R}$

b. From Table F.7.3 compressed liquid states at 500 psia, we get

$u_2 - u_1 = 373.68 - 167.64 = \mathbf{206.0\ Btu/lbm}$

$s_2 - s_1 = 0.5660 - 0.2934 = \mathbf{0.2726\ Btu/lbm\ R}$

c. From Table F.7.2 superheated vapor states at 10 psia, we get

$u_2 - u_1 = 1146.10 - 1074.67 = \mathbf{71.43\ Btu/lbm}$

$s_2 - s_1 = 1.9171 - 1.7927 = \mathbf{0.1244\ Btu/lbm\ R}$

d. From Table F.7.2 superheated vapor states at 1 psia, we get

$u_2 - u_1 = 1147.02 - 1077.49 = \mathbf{69.53\ Btu/lbm}$

$s_2 - s_1 = 2.1720 - 2.0507 = \mathbf{0.1213\ Btu/lbm\ R}$

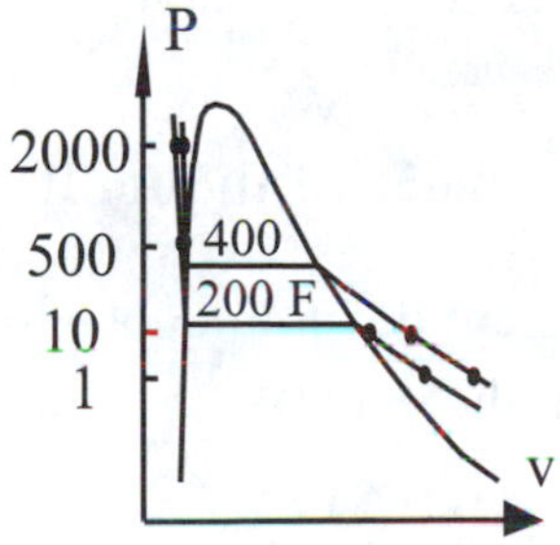

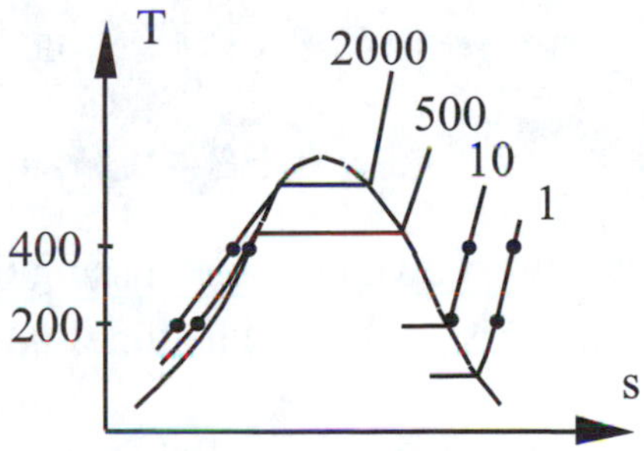

The changes in u and s are larger for the liquid than for the gas. Within the liquid phase the influence of P is very modest. Water vapor at 10 psia is close to an ideal gas ($Z = Pv_g/RT_{sat} = 0.988$). For the vapor phase changing P from 10 psia to 1 psia at a fixed T changes s, but u is nearly constant. The influence of P becomes much stronger as the state approaches the dense fluid region near the critical point.

8.4E An internally reversible, externally irreversible process

Ammonia in a piston-cylinder that maintains constant P is at 70 F, 250 psia and is now heated to 140 F in an internally reversible process. The external heat source is at a constant 160 F. We want to find the specific heat transfer in the process and the external (total) specific entropy generation.

Solution:

Take as a control volume the ammonia only, this is a control mass. Neglect storage of energy and entropy in the cylinder walls and piston mass.

Energy Eq.: $\quad u_2 - u_1 = {}_1q_2 - {}_1w_2$

Entropy Eq.: $\quad s_2 - s_1 = \int_1^2 \frac{dq}{T} + {}_1s_{2\text{ gen ammonia}}$

Process: $\quad P = C \;\Rightarrow\; {}_1w_2 = \int P\,dv = P\,(v_2 - v_1)$

$\quad$ Reversible $\;\Rightarrow\; {}_1s_{2\text{ gen ammonia}} = 0$

State 1: Table F.8.1 Approximate compressed liquid state with saturated same T

$$v_1 = 0.02631 \text{ ft}^3/\text{lbm}, \quad u_1 = 119.58 \text{ Btu/lbm}, \quad s_1 = 0.2529 \text{ Btu/lbm R}$$

State 2: Table F.8.2 Superheated vapor

$$v_2 = 1.3150 \text{ ft}^3/\text{lbm}, \quad s_2 = 1.1930 \text{ Btu/lbm R}$$

$$u_2 = 655.95 - 250 \times 1.315 \times 144 / 778 = 595.1 \text{ Btu/lbm}$$

Work from the process equation is

$$_1w_2 = P\,(v_2 - v_1) = 250 \text{ psia } (1.3150 - 0.02631) \text{ ft}^3/\text{lbm} = 59.63 \text{ Btu/lbm}$$

Heat transfer is from the energy equation (we could also have done it as $h_2 - h_1$)

$$_1q_2 = u_2 - u_1 + {}_1w_2 = 595.1 - 119.58 + 59.63 = \mathbf{535.2\ Btu/lbm}$$

The external entropy generation is a q over a ΔT so we can do it as in Eq.8.18:

$$_1s_{2\text{ gen tot}} = \Delta s_{CV} + \Delta s_{SUR} = s_2 - s_1 + \left[- \frac{_1q_2}{T_{source}} \right]$$

$$= 1.193 - 0.2529 - \frac{535.2}{459.67 + 160} = \mathbf{0.0764\ Btu/lbm\ R}$$

Choosing a CV as shown has $_1q_2$ crossing the CS at T_{source} not the ammonia T so the entropy equation becomes (and includes now s generated in the walls)

$$s_2 - s_1 = \int_1^2 \frac{dq}{T_S} + {}_1s_{2\text{ gen tot}} = \frac{_1q_2}{T_{source}} + {}_1s_{2\text{ gen tot}}$$

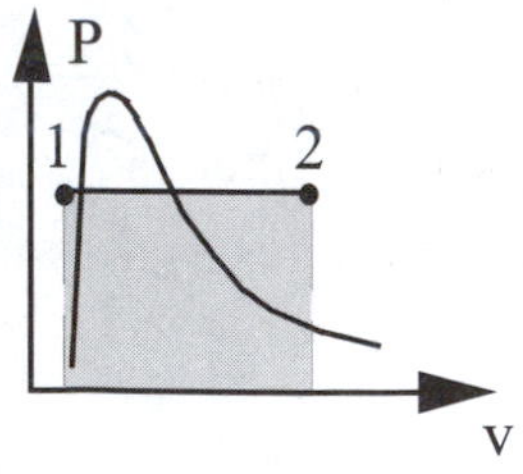

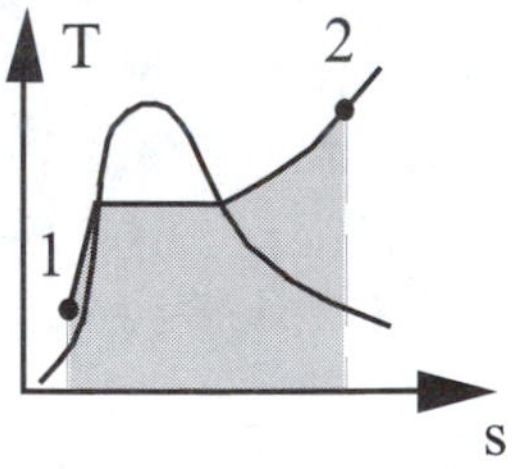

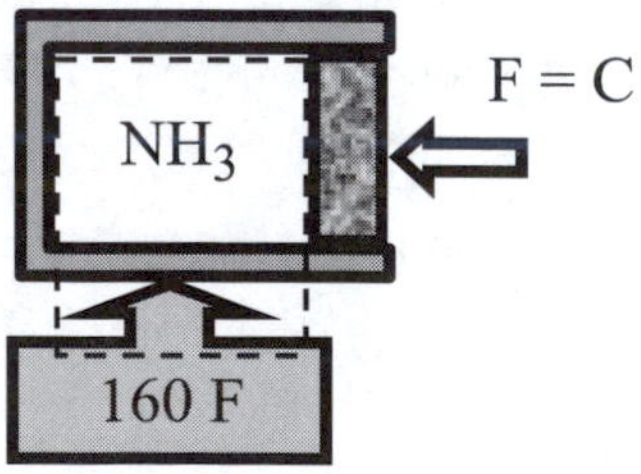

8.7E An isothermal expansion of air

A mass of 1 lbm of air contained in a cylinder at 200 psia, 1800 R, expands in a reversible isothermal process to a pressure 5 times smaller. Calculate the change of entropy of the air and the heat transfer during the process.

Solution:

C.V. Air, which is a control mass.

Energy Eq.5.11: $m(u_2 - u_1) = {}_1Q_2 - {}_1W_2$

Entropy Eq.8.14: $m(s_2 - s_1) = \int dQ/T + {}_1S_{2\ gen}$

Process: $T = \text{constant}$ so $\int dQ/T = {}_1Q_2 /T$

reversible implies ${}_1S_{2\ gen} = 0$

State 1: P_1, T_1 State 2: $T_2 = T_1,\ v_2 = v_1/5$

The change of entropy for the air (ideal gas) is from Eq.8.26

$$m(s_2 - s_1) = m[C_{vo} \ln \frac{T_2}{T_1} + R \ln \frac{v_2}{v_1}] = 1 [0 + \frac{53.34}{778} \ln(5)] = \mathbf{0.11\ Btu/R}$$

So the heat transfer from the second law becomes

$${}_1Q_2 = T\ m\ (s_2 - s_1) = 1800\ R \times 0.11\ Btu/R = \mathbf{198.6\ Btu}$$

Comment: For a non-ideal gas we would find state 2 in the tables and get (u_2, s_2), first do ${}_1Q_2$ and then do ${}_1W_2$ from the energy equation

$${}_1W_2 = {}_1Q_2 - m(u_2 - u_1)$$

For an ideal gas constant T => $u_2 = u_1$ then ${}_1W_2 = {}_1Q_2$

From the process equation and ideal gas law (PV = mRT = constant) we can calculate the work term (polytropic process) as in Eq.4.5

$${}_1Q_2 = {}_1W_2 = \int PdV = P_1V_1 \ln (V_2/V_1) = mRT_1 \ln (V_2/V_1)$$

$$= 1 \times \frac{53.34}{778} \times 1800 \ln (5) = 0.11\ Btu$$

consistent with the above result.

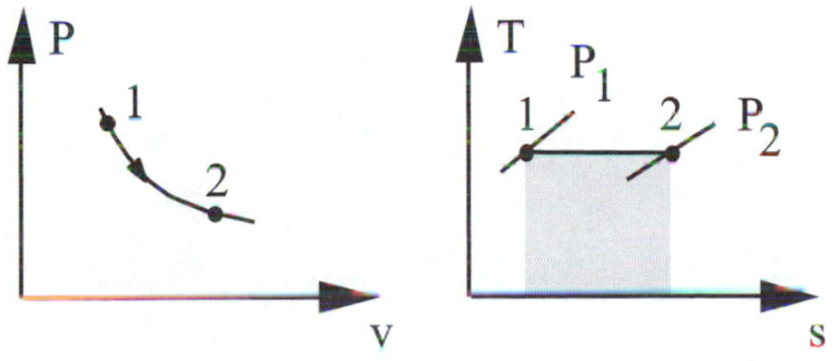

CHAPTER 9
STUDY PROBLEMS

SECOND-LAW ANALYSIS FOR A CONTROL VOLUME

- **The entropy equation for a control volume**
- **The steady state and the transient process**
- **The reversible steady state process**
- **Principle of the increase of entropy**
- **Efficiency**

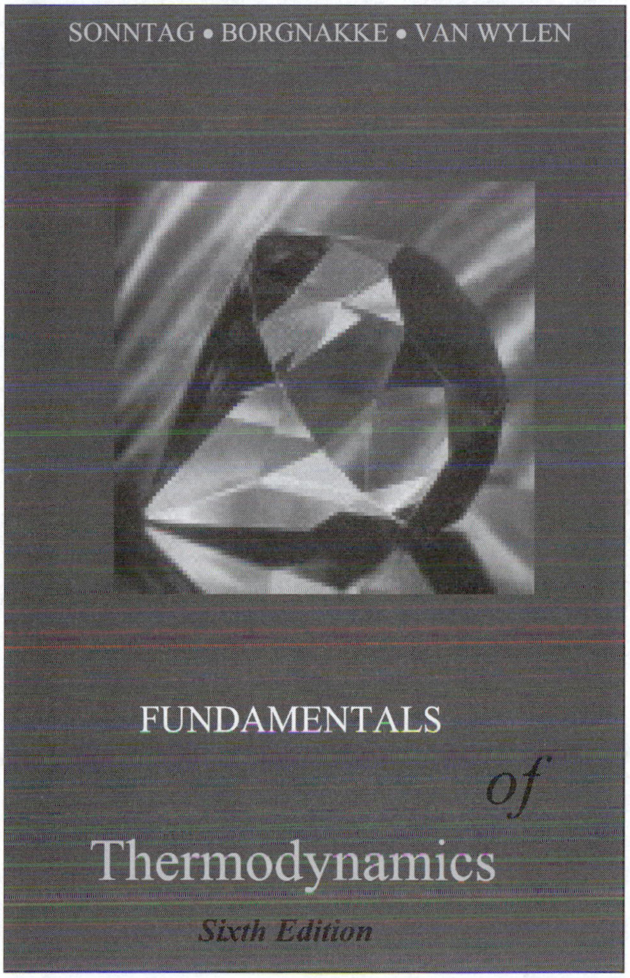

9.1 An ideal steam turbine

A steam turbine receives 4 kg/s steam at 1 MPa 300°C and there are two exit flows, 0.5 kg/s exits at 150 kPa and the rest exits at 15 kPa. Assume the turbine is ideal, adiabatic and that we can neglect kinetic energy everywhere. We want to determine the total power output.

Solution:

C.V. Turbine, steady operation adiabatic and reversible.

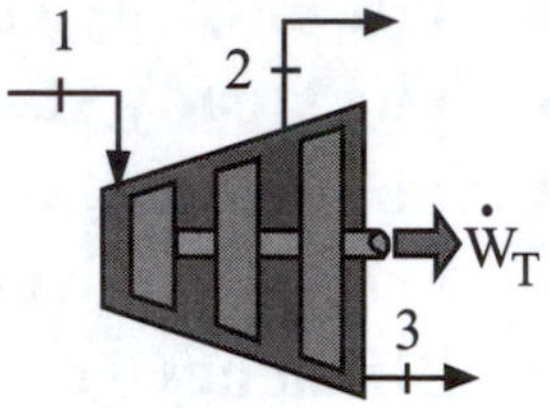

Continuity Eq.6.9: $\dot{m}_1 = \dot{m}_2 + \dot{m}_3$;

Energy Eq.6.10: $\dot{m}_1 h_1 = \dot{m}_2 h_2 + \dot{m}_3 h_3 + \dot{W}_T$

Entropy Eq.9.7: $\dot{m}_1 s_1 + \dot{S}_{gen} = \dot{m}_2 s_2 + \dot{m}_3 s_3$

Process: $\dot{S}_{gen} = 0$

If we separately applied the second law to section 1 to 2 and section 2 to 3 we get

Section 1 to 2: $\dot{m}_1 s_1 = \dot{m}_2 s_2 + \dot{m}_3 s_2 = \dot{m}_1 s_2 \quad \Rightarrow \quad s_1 = s_2$

Section 2 to 3: $\dot{m}_3 s_2 = \dot{m}_3 s_3 \quad \Rightarrow \quad s_2 = s_3$

$s_1 = s_2 = s_3$ that is constant s through turbine

State 1 from Table B.1.3: $h_1 = 3051.15$ kJ/kg, $s_1 = 7.1228$ kJ/kg K,

State 2 (P, s) : $s_2 < s_g$ so this state is two-phase

$$x_2 = (s_2 - s_f)/ s_{fg} = (7.1228 - 1.4335) / 5.7897 = 0.98266,$$

$$h_2 = 467.08 + x_2 \times 2226.46 = 2654.9 \text{ kJ/kg}$$

State 3 (P, s) : $s_3 < s_g$ so this state is two-phase

$$x_3 = (s_3 - s_f)/ s_{fg} = (7.1228 - 0.7548) / 7.2536 = 0.87791,$$

$$h_3 = 225.91 + x_3 \times 2373.14 = 2309.3 \text{ kJ/kg}$$

Now we can substitute into the energy equation

$$\dot{W}_T = \dot{m}_1 h_1 - \dot{m}_2 h_2 - \dot{m}_3 h_3$$

$$= 4 \times 3051.15 - 0.5 \times 2654.9 - 3.5 \times 2309.3 = \mathbf{2795\ kW}$$

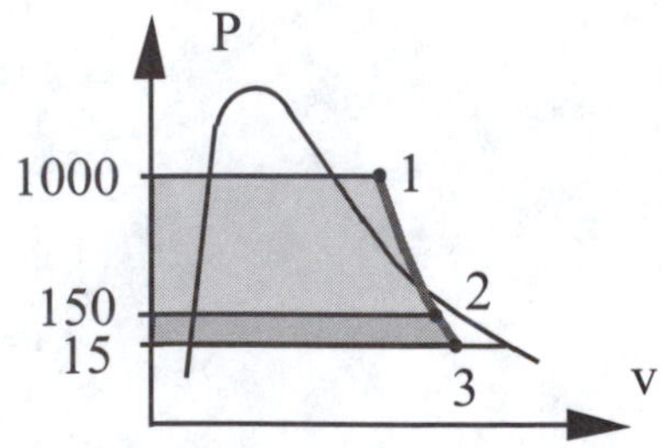

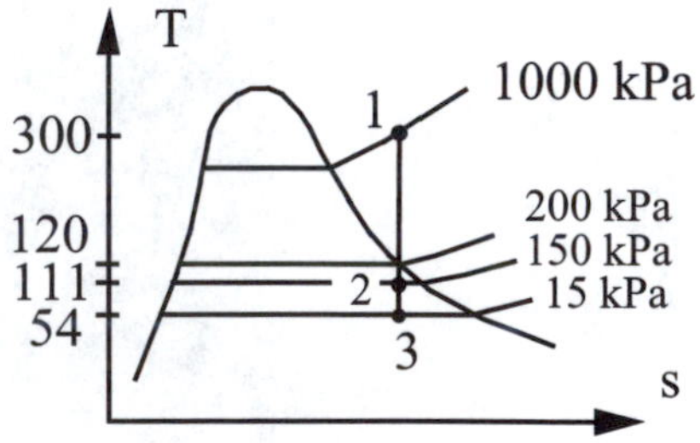

9.2 A cross-flowing heat exchanger with entropy generation

An ordinary radiator receives 0.2 kg/s hot water at 80°C, 110 kPa from a pipe and the water exits at 70°C. The radiator heats the air outside of it so the air rises up (natural convection) pulling colder air towards the radiator behind it. This acts as a two fluid heat exchanger so assume the air comes in at 20°C and leaves the radiator at 30°C. We want to know the mass flow rate of air and the total entropy generation rate in the process.

Solution:

The schematic may look like this:

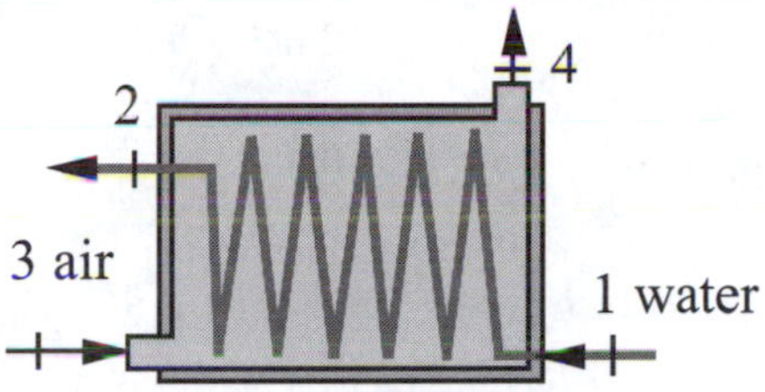

Energy Eq.6.10: $\dot{m}_{H2O}\, h_1 + \dot{m}_{AIR}\, h_3 = \dot{m}_{H2O}\, h_2 + \dot{m}_{AIR}\, h_4$

Entropy Eq.9.7: $\dot{m}_{H2O}\, s_1 + \dot{m}_{AIR}\, s_3 + \dot{S}_{gen} = \dot{m}_{H2O}\, s_2 + \dot{m}_{AIR}\, s_4$

Process: Constant pressure for air

From B.1.1: $h_1 = 334.88$ kJ/kg; $s_1 = 1.0752$ kJ/kg K

$h_2 = 292.96$ kJ/kg; $s_2 = 0.9548$ kJ/kg K

Using A.5: $h_4 - h_3 = C_p(T_4 - T_3) = 1.004(30 - 20) = 10.04$ kJ/kg

$$s_4 - s_3 = C_p \ln\left(\frac{T_4}{T_3}\right) - R \ln\left(\frac{P_4}{P_3}\right)$$

$$= 1.004 \ln\frac{30 + 273}{20 + 273} - 0 = 0.03368 \text{ kJ/kg K}$$

From energy equation

$$\dot{m}_{AIR} = \dot{m}_{H2O}\, \frac{h_1 - h_2}{h_4 - h_3} = 0.2\, \frac{334.88 - 292.96}{10.04} = \mathbf{0.835\ kg/s}$$

From entropy equation

$$\dot{S}_{gen} = \dot{m}_{H2O}(s_2 - s_1) + \dot{m}_{AIR}(s_4 - s_3)$$

$$= 0.2\,(0.9548 - 1.0752) + 0.835 \times 0.03368$$

$$= -\,0.02408 + 0.02812 = \mathbf{0.004\ kW/K}$$

9.3 Air flow in a jet engine exit nozzle

Air leaving the turbine section in a jet engine at 1100 K, 450 kPa with a velocity of 200 m/s enters a nozzle with an exit state of 100 kPa. Find the exit temperature, the exit velocity and the ratio of the exit area to the nozzle inlet area A_{ex}/A_{in} in case of a reversible adiabatic flow.

Solution:

C.V. around the nozzle section, this is steady state, single flow with no heat transfer and no work, also same elevation $Z_{in} = Z_{ex}$.

Continuity Eq.: $\dot{m}_{in} = \dot{m}_{ex} = \dot{m} = A_{in}\mathbf{V}_{in}/v_{in} = A_{ex}\mathbf{V}_{ex}/v_{ex}$

Energy Eq.6.12: $\dot{m}(h_{in} + \frac{1}{2}\mathbf{V}_{in}^2) = \dot{m}(h_{ex} + \frac{1}{2}\mathbf{V}_{ex}^2)$

Entropy Eq. 9.8: $\dot{m}s_{in} + \frac{\dot{Q}}{T} + \dot{S}_{gen} = \dot{m}s_{in} + 0 + 0 = \dot{m}s_{ex}$

State properties from Table A.7.1: $h_{in} = 1161.18$ kJ/kg, $s^o_{T\,in} = 8.24449$ kJ/kg-K.

From the entropy equation we have constant s implemented as in Eq. 8.28 gives

$$s^o_{T\,ex} = s^o_{T\,in} + R \ln\left(\frac{P_{ex}}{P_{in}}\right) = 8.24449 + 0.287 \ln\left(\frac{100}{450}\right) = 7.81282 \text{ kJ/kg-K}$$

Now back-interpolate in Table A.7.1 between 740 K and 760 K to get

$$T_{ex} = 748.8 \text{ K} \quad \text{and} \quad h_{ex} = 766.286 \text{ kJ/kg}$$

From the energy equation solve for the exit kinetic energy, notice conversion from kJ to J for the enthalpy terms

$$\frac{1}{2}\mathbf{V}_{ex}^2 = \frac{1}{2}\mathbf{V}_{in}^2 + h_{in} - h_{ex} = \frac{1}{2}200^2 + (1161.18 - 766.286)\,1000$$

$$= 20\,000 + 394\,894 = 414\,894 \text{ m}^2/\text{s}^2$$

$$\mathbf{V}_{ex} = \sqrt{2 \times 414\,894} = \mathbf{910.9 \text{ m/s}}$$

From the continuity equation and ideal gas law, $v = RT/P$, we have

$$A_{ex} / A_{in} = \frac{\mathbf{V}_{in}}{\mathbf{V}_{ex}} \times \frac{T_{ex}}{P_{ex}} \times \frac{P_{in}}{T_{in}} = \frac{200}{910.9} \times \frac{748.8}{1100} \times \frac{450}{100} = \mathbf{0.673}$$

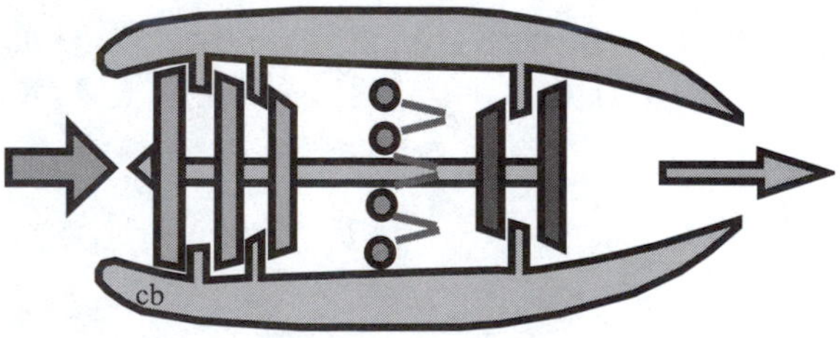

9.4 A steady flow mixing process

A flow of 1 kg/s air at 300 K, 1000 kPa is throttled down to 500 kPa and mixed with another flow of 2 kg/s air at 1000 K, 500 kPa in an open mixing chamber producing a uniform exit flow at 500 kPa. What is the rate of entropy generation in this process?

Solution:

CV. Throttle (valve) and the mixing chamber.

Continuity Eq.6.9: $\dot{m}_1 + \dot{m}_2 = \dot{m}_3 = 1 + 2 = 3$ kg/s

Energy Eq.6.10: $\dot{m}_1 h_1 + \dot{m}_2 h_2 = \dot{m}_3 h_3$

Entropy Eq.9.7: $\dot{m}_1 s_1 + \dot{m}_2 s_2 + \dot{S}_{gen} = \dot{m}_3 s_3$

Let us solve using constant specific heats from A.5 and Eq.8.25 for s change.
Divide the energy equation with $\dot{m}_3 C_{Po}$

$$T_3 = (\dot{m}_1/\dot{m}_3)T_1 + (\dot{m}_2/\dot{m}_3)T_2 = \frac{1}{3} \times 300 + \frac{2}{3} \times 1000 = 766.67 \text{ K}$$

$$\dot{S}_{gen} = \dot{m}_1(s_3 - s_1) + \dot{m}_2(s_3 - s_2)$$

$$= 1 \times [1.004 \ln(\frac{766.67}{300}) - 0.287 \ln(\frac{500}{1000})] + 2 \times [1.004 \ln(\frac{766.67}{1000}) - 0]$$

$$= \mathbf{0.607 \ kW/K}$$

Solve now (more accurately) using Table A.7.1 and Eq.8.28 for change in s.

$$h_3 = (\dot{m}_1/\dot{m}_3)h_1 + (\dot{m}_2/\dot{m}_3)h_2 = \frac{1}{3} \times 300.47 + \frac{2}{3} \times 1046.22 = 797.637 \text{ kJ/kg}$$

From A.7.1: $T_3 = 777.58$ K $\qquad s^o_{T3} = 7.85396$ kJ/kg K

$$\dot{S}_{gen} = 1[7.85396 - 6.86926 - 0.287 \ln(500/1000)] + 2(7.85396 - 8.13493)$$

$$= \mathbf{0.622 \ kW/K}$$

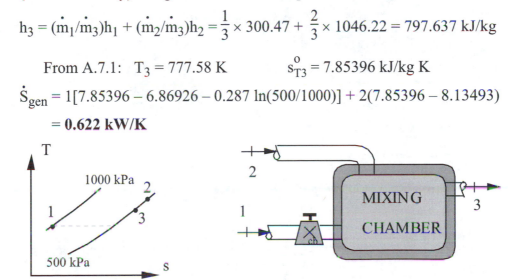

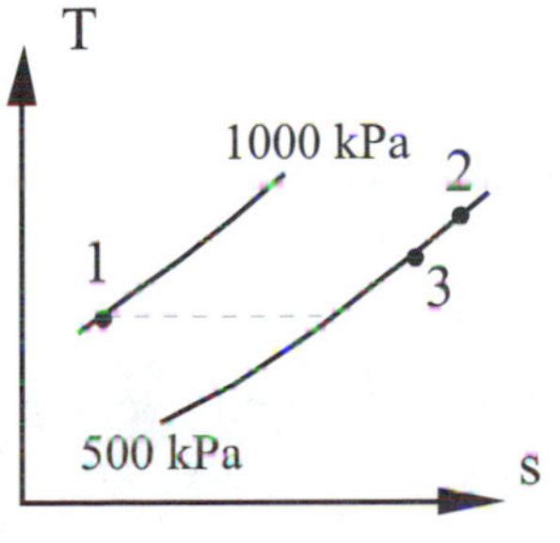

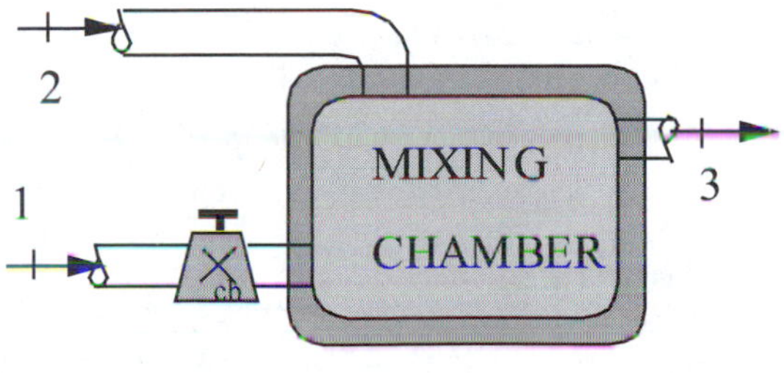

9.5 Flow of liquid through a tank and nozzle

A tall tank of cross section area 5 m^2 receives a flow of 50 kg/s water at 101 kPa, 20°C with negligible kinetic energy. At the bottom of the tank is a nozzle pointing downwards with the smallest diameter of 0.1 m. Assume a reversible flow with no change in temperature and find the exit velocity and the water level height H that gives steady state in the tank.

Solution:

C.V.: Tank and its water. No work or heat transfer. To have steady state the mass flow rate out must be the same as the inlet mass flow rate.

Continuity Eq.6.3, 6.11: $\dot{m}_{in} = \dot{m}_{ex} = (\rho A \mathbf{V})_{nozzle}$

Energy Eq.6.12: $\dot{m}(h + \mathbf{V}^2/2 + gZ)_{in} = \dot{m}(h + \mathbf{V}^2/2 + gZ)_{ex}$

Entropy Eq.9.8: $\dot{m}s_{in} + \dot{S}_{gen} = \dot{m}s_{ex}$

Process: Reversible and $\mathbf{V}_{in} \cong 0$, $Z_{in} - Z_{ex} = H$, $\rho = 1/v \cong 1/v_f$

For the reversible process the second law leads to Eq.9.13

$$h_{ex} - h_{in} = \int_i^e v\, dP = 0 \quad \text{(same P and constant v)}$$

With zero work the energy equation therefore becomes Bernoulli Eq.9.17

$$g(Z_{in} - Z_{ex}) = g\,H = \mathbf{V}_{ex}^2/2$$

From the continuity equation

$$\mathbf{V}_{ex} = \frac{\dot{m}_{in}}{\rho A_{nozzle}} = \frac{4\dot{m}_{in}v}{\pi D^2} = \frac{4 \times 50 \text{ kg/s} \times 0.001002 \text{ m}^3\text{/kg}}{\pi\, 0.1^2 \text{ m}^2} = \mathbf{6.379\ m/s}$$

From Bernoulli's equation we then get

$$H = \mathbf{V}_{ex}^2/2g = \frac{(6.379 \text{ m/s})^2}{2 \times 9.81 \text{ m/s}^2} = \mathbf{2.074\ m}$$

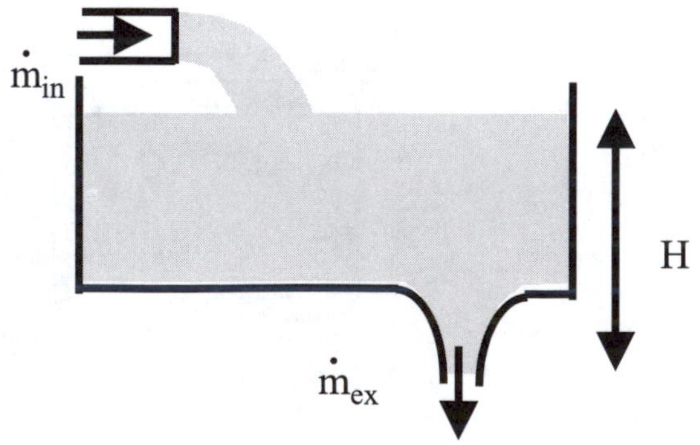

9.6 Stagnation point of a jet flow

In a sand blast application we want to direct a jet flow towards a surface where the flow stagnates. The stagnation pressure should be at least 200 kPa. Find the velocity we need to create and the pump/compressor work required if we use a. liquid water or b. air. Both flows would be taken in at atmospheric conditions 100 kPa, 17°C.

Solution:

C.V. The pump/compressor, the pipe, the nozzle and the jet flow to the stagnation point. Steady state flow with no elevation changes and we will assume no heat transfer. We also will neglect the kinetic energy in the inlet flow and in the pipe flow and by definition the stagnation velocity is zero.

Process: $\mathbf{V}_1 = \mathbf{V}_2 = \mathbf{V}_4 = 0$, $\Delta Z = 0$, assume reversible flow $s_{gen} = 0$

Energy Eq.6.13: $h_1 + w_{in} = h_2 = h_3 + \frac{1}{2}\mathbf{V}_3^2 = h_4$

Entropy Eq.9.8: $s_1 = s_2 = s_3 = s_4$ (Eq. is divided by $\dot{m}$)

States: 1: (T_1, P_1) 2: (h_4, s_1) 3: (P_1, s_1) 4: (P_{stag}, s_1)

We see that state 2 equals state 4 (same h and s), state 3 equals state 1, but with a $\mathbf{V}_3$

Energy Eq. and Eq.9.13 $w_{in} = h_4 - h_1 = \frac{1}{2}\mathbf{V}_3^2 = h_4 - h_3 = \int_3^4 v\,dP$

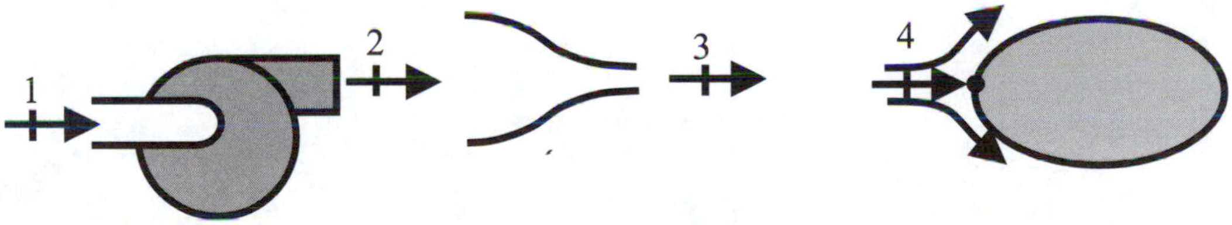

a) Water:

From Table A.4 (or B.1.1) $\rho = 997\ kg/m^3 = 1/v$; **incompressible**

$$w_{in} = \int_3^4 v\,dP = v\,(P_4 - P_3) = (200 - 100) / 997 = \mathbf{0.1003\ kJ/kg}$$

$$\mathbf{V}_3 = \sqrt{2w_{in}} = \sqrt{2\Delta P/\rho} = \sqrt{2000 \times 0.1003\ J/kg} = \mathbf{14.2\ m/s}$$

Remark: This is rather low velocity for a water jet and an actual jet would use a much higher velocity and give a higher stagnation pressure.

b) **Air:**

Ideal gas: $v_1 = RT_1/P_1 = 0.287 \times 290/100 = 0.8323 \text{ m}^3/\text{kg}$

Let us assume the flow is incompressible and do it as for water

$$w_{in} = \int_3^4 v\, dP = v\,(P_4 - P_3) = 0.8323\,(200 - 100) = \mathbf{83.23\ kJ/kg}$$

$$V_3 = \sqrt{2w_{in}} = \sqrt{2\Delta P/\Box} = \mathbf{408\ m/s}$$

This is high velocity (120% of the speed of sound of about 340 m/s)

Analyze with air as a compressible substance and find state 4: (P_{stag}, s_1)

Eq.8.32: $$T_4 = T_1 \left[P_{stag} / P_1\right]^{(k-1)/k} = 290.15 \left[200 / 100\right]^{0.2857} = 353.69 \text{ K}$$

Now we get from the energy equation:

$$w_{in} = h_4 - h_1 = C_p\,(T_4 - T_1) = 1.004\,(353.69 - 290.15) = \mathbf{63.79\ kJ/kg}$$

and

$$V_3 = \sqrt{2w_{in}} = \sqrt{2000 \times 63.79} = \mathbf{357\ m/s}$$

Remark: Assuming incompressible flow is somewhat inaccurate as the specific volume used for calculating work is overestimated.

9.7 A steady flow pump and a filling process

Assume we have a 4 m tall storage tank, initially empty. A submerged pump in an open well 10 m down fills the tank with water from the well at 10°C and 100 kPa. The pump exit is into a pipe that goes up to the top of the storage tank where it just splashes out into the tank. Neglect the kinetic energy of the flow and assume we fill the storage tank so it has 1 m of liquid. How much work per unit mass was used to drive the pump (assume reversible i.e. 100% efficient). Find the final temperature of the stored water (no heat transfer) and the specific entropy generated in the process.

In this process we should realize the pump operates in a reversible steady flow process whereas the tank goes through a transient process.

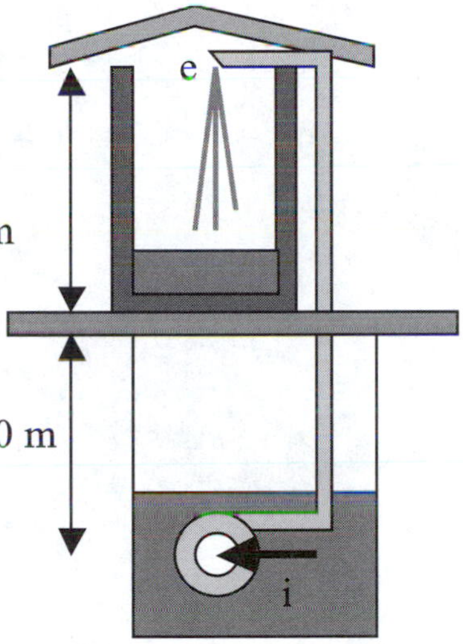

Solution:

CV. Pump and pipe ($q = 0$ and $s_{gen} = 0$).

Energy Eq.6.13: $h_i + w_{p\ in} + gZ_i = h_e + gZ_e$

Entropy Eq.9.8: $s_e - s_i = \int_i^e \frac{dq}{T} + s_{gen} = 0$

So constant s and same P = 100 kPa in and out give the same T (recall Eq. 8.20) so

$$h_i = h_e$$

This gives

$$w_{p\ in} = g(Z_e - Z_i) = 9.807\ \text{ms}^{-2} \times (4 + 10)\ \text{m} = \mathbf{137.3\ J/kg}$$

C.V. Water in the tank (volume expands against P_o), inlet state equal to pipe exit state.

Continuity Eq.: $m_2 - m_1 = m_i \quad => \quad m_2 = m_i = V_2/v_2$

Energy Eq.: $m_2(u_2 + gZ_2) - 0 = m_i(h_i + gZ_i) - W_o = m_i(h_i + gZ_i) - P_oV_2$

Entropy Eq.9.12: $m_2s_2 - 0 = m_is_i + 0 + {}_1S_{2\ gen}$

From the energy equation we get after dividing m_2 out:

$$u_2 + gZ_2 + P_ov_2 = h_i + gZ_i \quad => \quad h_2 = h_i + g(Z_i - Z_2)$$

$$T_2 = T_i + g(Z_i - Z_2)/C_p = 10 + 9.807\ (4 - 0.5) / 4180 = 10.0082^\circ\text{C}$$

From the entropy equation we get (recall Eq.8.20)

$${}_1s_{2\ gen} = (s_2 - s_i) = C\ \ln(T_2/T_i) = 4180\ \ln\frac{283.1582}{283.15} = \mathbf{0.12\ J/kg\ K}$$

Comment: The irreversible part is conversion of potential energy (top of the tank) into a lower potential energy (average $Z_2 = 0.5$ m). Of course the magnitude of the elevation differences makes the temperature rize and entropy generation virtually zero.

9.8 A polytropic expansion process

In a planned hydrogen storage facility an expander (turbine with heat transfer) brings 0.175 kg/s hydrogen gas from 1000 kPa, 550°C to 200 kPa. Assume the process is a polytropic process with n = 1.5. What are the work and heat transfer in the expander?

Solution:

CV: expander, steady single inlet and single exit flow with both $\dot{W}$ and $\dot{Q}$.

Energy Eq.6.13: $h_i + q = h_e + w$

Entropy Eq.9.8: $s_e - s_i = \int_i^e \frac{dq}{T} + s_{gen}$

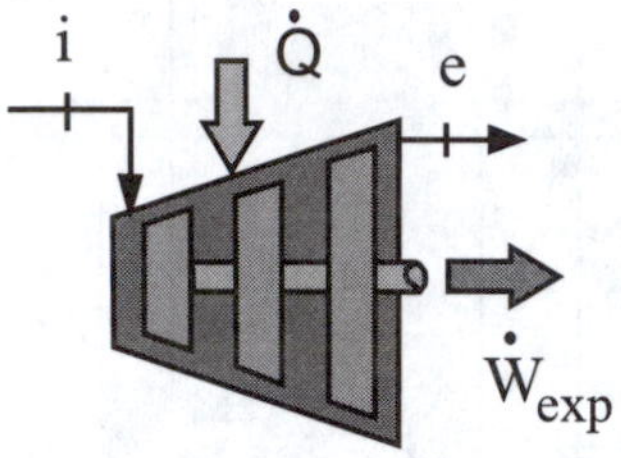

Reversible polytropic process from Eq.8.37:

$$T_e = T_i \left(\frac{P_e}{P_i}\right)^{\frac{n-1}{n}} = 823.2 \left(\frac{200}{1000}\right)^{\frac{0.5}{1.5}} = 481 \text{ K}$$

Work evaluated from Eq.9.19

$$w = -\int v dP = -\frac{nR}{n-1}(T_e - T_i) = \frac{-1.5 \times 4.1243}{0.5}(481 - 823.2) = 4231 \text{ kJ/kg}$$

$$\dot{W} = \dot{m}w = 0.175 \times 4231 = \mathbf{740\ kW}$$

From the energy equation we have

$$q = h_e - h_i + w = C_p (T_e - T_i) + w$$

$$= 14.21 (481 - 823.2) + 4231 = -628.3 \text{ kJ/kg}$$

$$\dot{Q} = \dot{m}q = 0.175 \times (-628.3) = \mathbf{-110\ kW}$$

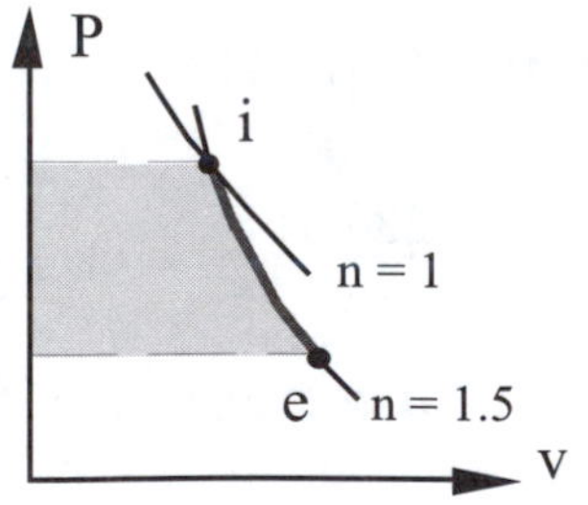

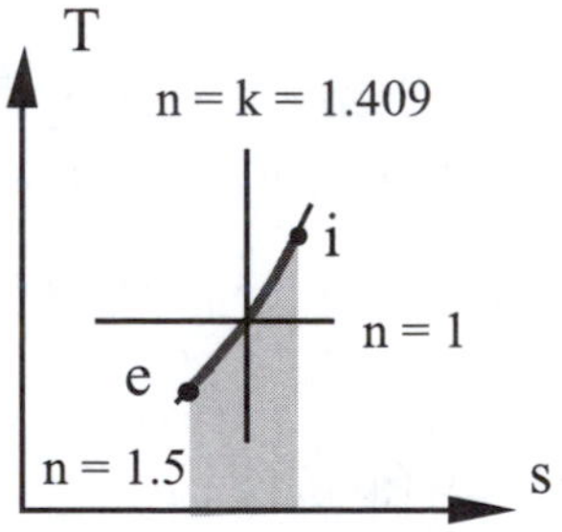

9.9 The efficiency of an actual compressor

The compressor in example 6.7, page 178, receives carbon dioxide at 100 kPa, 280 K, with a low velocity. At the compressor discharge, the carbon dioxide exits at 1100 kPa, 500 K, with a velocity of 25 m/s. The actual power input to the compressor is 50 kW. We want to find the isentropic efficiency and the total rate of entropy generation.

Solution:

C.V. Actual compressor, steady state, single inlet and exit flow.

Energy Eq.6.13: $q + h_1 + \frac{1}{2}\mathbf{V}_1^2 = h_2 + \frac{1}{2}\mathbf{V}_2^2 + w$

Entropy Eq.9.8: $\dot{m}s_1 + \dot{S}_{gen} = \dot{m}s_2$

Here we assume $q \cong 0$ and $\mathbf{V}_1 \cong 0$ so getting h from Table A.8

$$-w = h_2 - h_1 + \frac{1}{2}\mathbf{V}_2^2 = 401.52 - 198 + \frac{(25)^2}{2 \times 1000} = 203.5 + 0.3 = 203.8 \text{ kJ/kg}$$

remember here to convert kinetic energy J/kg to kJ/kg by division with 1000.

$$\dot{m} = \frac{\dot{W}_c}{w} = \frac{-50}{-203.8} = 0.245 \text{ kg/s}$$

Solve for $\dot{S}_{gen}$ and use Eq.8.28 with the standard entropy from Table A.8

$$\dot{S}_{gen} = \dot{m}\,(s_2 - s_1) = \dot{m}\,[s^o_{T2} - s^o_{T1} - R \ln(P_2/P_1)]$$

$$= 0.245\,[\,5.3375 - 4.8034 - 0.1889 \ln(11)\,]$$

$$= \mathbf{0.02\ kW/K}$$

C.V. Ideal (reversible) adiabatic compressor, exit state (2s) at 1100 kPa, but not 500 K.

Energy Eq.6.13: $h_1 + 0 = h_{2s} + \frac{1}{2}\mathbf{V}_2^2 + w_s$

Entropy Eq.9.8: $s_1 + 0 = s_{2s}$ (Eq. divided by mass flow rate)

Ideal exit state 2s: $(P_2, s = s_1)$ same s as state 1 hence the name 2s

Eq.8.28: $s^o_{T2s} = s^o_{T1} + R \ln(\frac{P_2}{P_1}) = 4.8034 + 0.1889 \ln(11) = 5.2564$ kJ/kgK

Interpolate in Table A.8: $T_{2s} = 461.4$ K and $h_{2s} = 363$ kJ/kg

Work from the energy equation becomes

$$-w_s = h_{2s} + \frac{1}{2}\mathbf{V}_2^2 - h_1 = 363 + 0.3 - 198 = 165.3 \text{ kJ/kg}$$

Efficiency, Eq.9.28: $\eta = \frac{w_s}{w} = \frac{165.3}{203.5} = \mathbf{0.81}$

9.10 The efficiency of a real nozzle

The real nozzle in study problem 6.4 was redone as an ideal nozzle in study problem 9.3. By comparing the two find the nozzle efficiency.

Air leaving the turbine section in a jet engine at 1100 K, 450 kPa with a velocity of 200 m/s enters a nozzle with an exit state of 100 kPa and 780 K. Find the exit velocity and the ratio of the exit area to the nozzle inlet area A_{ex}/A_{in}.

From the energy equation for the actual nozzle, see study problem 6.4, solve for the exit kinetic energy, notice conversion from kJ to J for the enthalpy terms (from A.7.1)

$$\frac{1}{2}\mathbf{V}^2_{ex\ ac} = \frac{1}{2}\mathbf{V}^2_{in} + h_{in} - h_{ex\ ac} = \frac{1}{2}\,200^2 + (1161.18 - 800.28) \times 1000$$

$$= 20\,000 + 360\,900 = 380\,900 \text{ m}^2/\text{s}^2$$

Now the ideal nozzle has a different exit state (same P but a lower T) and the analysis in study problem 9.3 gave the condition as (P_{ex}, $s_{ex} = s_{in}$). Therefore

$$\frac{1}{2}\mathbf{V}^2_{ex\ s} = \frac{1}{2}\mathbf{V}^2_{in} + h_{in} - h_{ex\ s} = \frac{1}{2}\,200^2 + (1161.18 - 766.286) \times 1000$$

$$= 20\,000 + 394\,894 = 414\,894 \text{ m}^2/\text{s}^2$$

The ratio of the two kinetic energies is the efficiency, Eq. 9.30

$$\eta = \frac{\frac{1}{2}\mathbf{V}^2_{ex\ ac}}{\frac{1}{2}\mathbf{V}^2_{ex\ s}} = \frac{380\,900}{414\,894} = \mathbf{0.92}$$

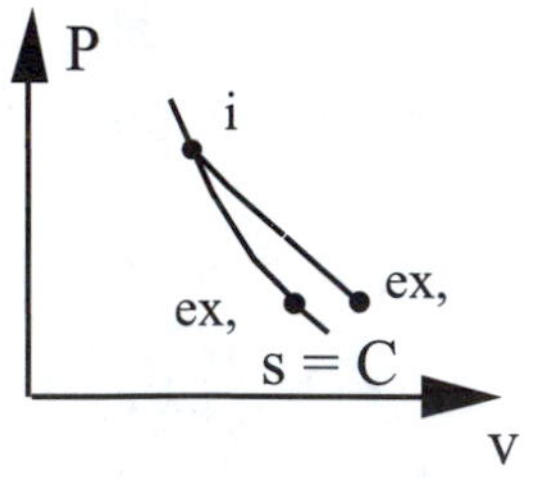

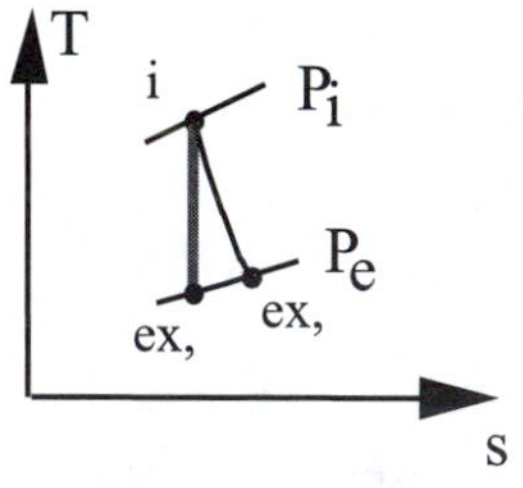

9.11 A mathematical solution to an unsteady problem

The water tank in study problem 9.5 has at one point a water level of 2.5 m when the inlet water flow is turned off. Estimate the time it will take to empty the tank.

Solution:

To solve the problem we neglect the volume and height of the nozzle so the tank volume is the tank cross sectional area times the height.

Continuity Eq.: $$\frac{dm_{tank}}{dt} = -\dot{m}_{ex} = -(\rho A\mathbf{V})_{nozzle}$$

$$m_{tank} = \rho\, V = \rho A_{tank}\, H$$

Energy equation is the same as in study problem 9.5, Bernoulli's Eq.9.17:

$$g(Z_{in} - Z_{ex}) = g\, H = \mathbf{V}_{ex}^2/2 \qquad \Rightarrow \qquad \mathbf{V}_{nozzle} = \sqrt{2gH}$$

$$\frac{d}{dt}[\rho A_{tank} H] = \rho A_{tank} \frac{dH}{dt} = -\rho A_{nozzle}\sqrt{2gH}$$

$$\frac{dH}{dt} = -\frac{A_{nozzle}}{A_{tank}}\sqrt{2gH} = -C\, H^{1/2}$$

Separate the variables

$$H^{-1/2}\, dH = -C\, dt$$

Now integrate the equation from the beginning to the end of the process

$$2H^{1/2}\Big|_{beg}^{end} = 2H_2^{1/2} - 2H_1^{1/2} = -C\,(t_2 - t_1)$$

To empty the tank we set the height at the end $H_2 = 0$ and solve for time

$$\Delta t = t_2 - t_1 = \frac{2\sqrt{H_1}}{C} = \frac{2\sqrt{H_1}\, A_{tank}}{A_{nozzle}\sqrt{2g}} = \left(\frac{A_{tank}}{A_{nozzle}}\right)\sqrt{\frac{2H_1}{g}}$$

$$\Delta t = \left(\frac{5}{0.007854}\right)\sqrt{\frac{2 \times 2.5}{9.81}} = \mathbf{454.5\ s = 7.6\ min}$$

9.2E A cross-flowing heat exchanger with entropy generation

An ordinary radiator receives 0.4 lbm/s hot water at 180 F, 16 psia from a pipe and the water exits at 160 F. The radiator heats the air outside of it so the air rises up (natural convection) pulling colder air towards the radiator behind it. This acts as a two fluid heat exchanger so assume the air comes in at 70 F and leaves the radiator at 86 F. We want to know the mass flow rate of air and the total entropy generation rate in the process.

Solution:

The schematic may look like this:

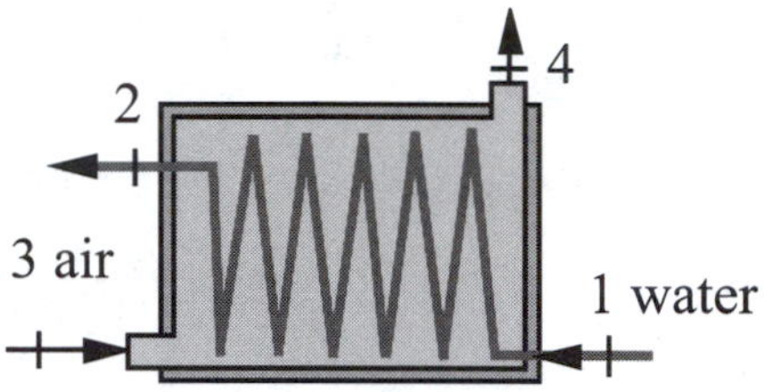

Energy Eq.6.10: $\dot{m}_{H2O}\, h_1 + \dot{m}_{AIR}\, h_3 = \dot{m}_{H2O}\, h_2 + \dot{m}_{AIR}\, h_4$

Entropy Eq.9.7: $\dot{m}_{H2O}\, s_1 + \dot{m}_{AIR}\, s_3 + \dot{S}_{gen} = \dot{m}_{H2O}\, s_2 + \dot{m}_{AIR}\, s_4$

Process: Constant pressure for air

From F.7.1: $h_1 = 147.98$ Btu/lbm; $s_1 = 0.2631$ Btu/lbm R

$h_2 = 127.95$ Btu/lbm; $s_2 = 0.2313$ Btu/lbm R

Using F.4: $h_4 - h_3 = C_p(T_4 - T_3) = 0.24(86 - 70) = 3.84$ Btu/lbm

$$s_4 - s_3 = C_p \ln\left(\frac{T_4}{T_3}\right) - R \ln\left(\frac{P_4}{P_3}\right) =$$

$$= 0.24 \ln \frac{86 + 459.67}{70 + 459.67} - 0 = 0.00714 \text{ Btu/lbm R}$$

From energy equation

$$\dot{m}_{AIR} = \dot{m}_{H2O} \frac{h_1 - h_2}{h_4 - h_3} = 0.4 \; \frac{147.98 - 127.95}{3.84} = \mathbf{2.086\ lbm/s}$$

From entropy equation

$$\dot{S}_{gen} = \dot{m}_{H2O}(s_2 - s_1) + \dot{m}_{AIR}(s_4 - s_3)$$

$$= 0.4\,(0.2313 - 0.2631) + 2.086 \times 0.00714$$

$$= -\,0.01272 + 0.01489 = \mathbf{0.0022\ Btu/s\text{-}R}$$

9.4E A steady flow mixing process

A flow of 1 lbm/s air at 540 R, 150 psia is throttled down to 75 psia and mixed with another flow of 2 lbm/s air at 1800 R, 75 psia in an open mixing chamber producing a uniform exit flow at 75 psia. What is the rate of entropy generation in this process?

Solution:

CV. Throttle (valve) and the mixing chamber.

Continuity Eq.6.9: $\dot{m}_1 + \dot{m}_2 = \dot{m}_3 = 1 + 2 = 3$ lbm/s

Energy Eq.6.10: $\dot{m}_1 h_1 + \dot{m}_2 h_2 = \dot{m}_3 h_3$

Entropy Eq.9.7: $\dot{m}_1 s_1 + \dot{m}_2 s_2 + \dot{S}_{gen} = \dot{m}_3 s_3$

Let us solve using constant specific heats from F.4 and Eq.8.25 for s change.

Divide the energy equation with $\dot{m}_3 C_{Po}$

$$T_3 = (\dot{m}_1/\dot{m}_3)T_1 + (\dot{m}_2/\dot{m}_3)T_2 = \frac{1}{3} \times 540 + \frac{2}{3} \times 1800 = 1380 \text{ R}$$

$$\dot{S}_{gen} = \dot{m}_1(s_3 - s_1) + \dot{m}_2(s_3 - s_2)$$

$$= 1 \times [0.24 \ln(\frac{1380}{540}) - \frac{53.34}{778} \ln(\frac{75}{150})] + 2 \times [0.24 \ln(\frac{1380}{1800}) - 0]$$

$$= \mathbf{0.145 \text{ Btu/s-R}}$$

Solve now (more accurately) using Table F.5 and Eq.8.28 for change in s.

$$h_3 = (\dot{m}_1/\dot{m}_3)h_1 + (\dot{m}_2/\dot{m}_3)h_2 = \frac{1}{3} \times 129.18 + \frac{2}{3} \times 449.794 = 342.92 \text{ Btu/lbm}$$

From F.5: $T_3 = 1399.6$ R $\quad s^o_{T3} = 1.875$ Btu/lbm R

$$\dot{S}_{gen} = 1[1.875 - 1.63979 - \frac{53.34}{778} \ln(75/150)] + 2(1.875 - 1.94209)$$

$$= \mathbf{0.148 \text{ Btu/s-R}}$$

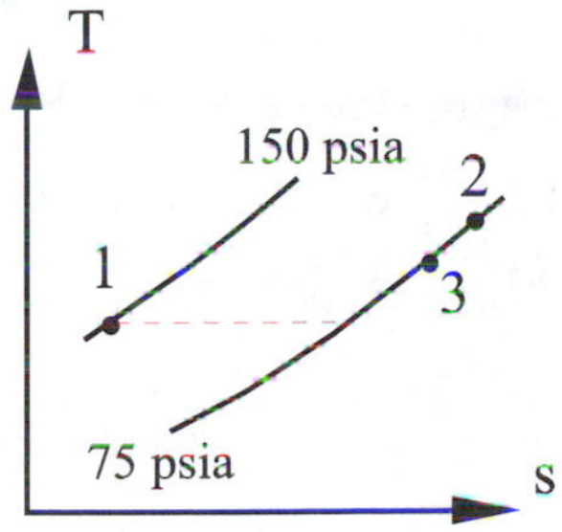

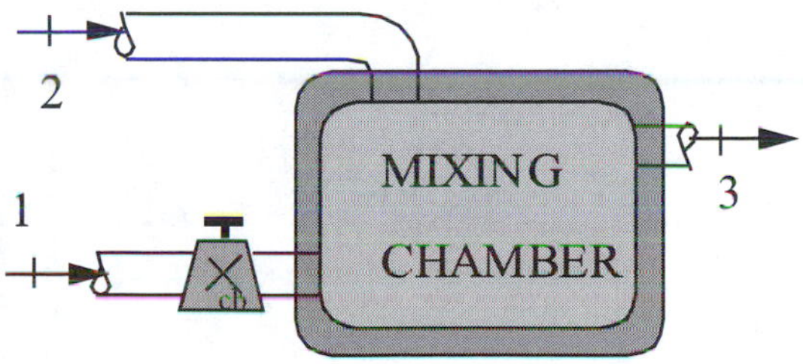

9.5E Flow of liquid through a tank and nozzle

A tall tank of cross section area 50 ft^2 receives a flow of 50 lbm/s water at 14.7 psia, 70 F with negligible kinetic energy. At the bottom of the tank is a nozzle pointing downwards with the smallest diameter of 3 in. Assume a reversible flow with no change in temperature and find the exit velocity and the water level height H that gives steady state in the tank.

Solution:

C.V.: Tank and its water. No work or heat transfer. To have steady state the mass flow rate out must be the same as the inlet mass flow rate.

Continuity Eq.6.3, 6.11: $\dot{m}_{in} = \dot{m}_{ex} = (\rho A \mathbf{V})_{nozzle}$

Energy Eq.6.12: $\dot{m}(h + \mathbf{V}^2/2 + gZ)_{in} = \dot{m}(h + \mathbf{V}^2/2 + gZ)_{ex}$

Entropy Eq.9.8: $\dot{m}s_{in} + \dot{S}_{gen} = \dot{m}s_{ex}$

Process: Reversible and $\mathbf{V}_{in} \cong 0$, $Z_{in} - Z_{ex} = H$, $\rho = 1/v \cong 1/v_f$

For the reversible process the second law leads to Eq.9.13

$$h_{ex} - h_{in} = \int_i^e v\, dP = 0 \quad \text{(same P and constant v)}$$

With zero work the energy equation therefore becomes Bernoulli Eq.9.17

$$g(Z_{in} - Z_{ex}) = g\,H = \mathbf{V}_{ex}^2/2$$

From the continuity equation

$$\mathbf{V}_{ex} = \frac{\dot{m}_{in}}{\rho A_{nozzle}} = \frac{4\dot{m}_{in} v}{\pi D^2} = \frac{4 \times 50 \text{ lbm/s} \times 0.01605 \text{ ft}^3\text{/lbm}}{\pi\, 3^2 \text{ in.}^2} = \mathbf{16.35\ ft/s}$$

From Bernoulli's equation we then get

$$H = \mathbf{V}_{ex}^2/2g = \frac{(16.35 \text{ ft/s})^2}{2 \times 32.174 \text{ ft/s}^2} = \mathbf{4.15\ ft}$$

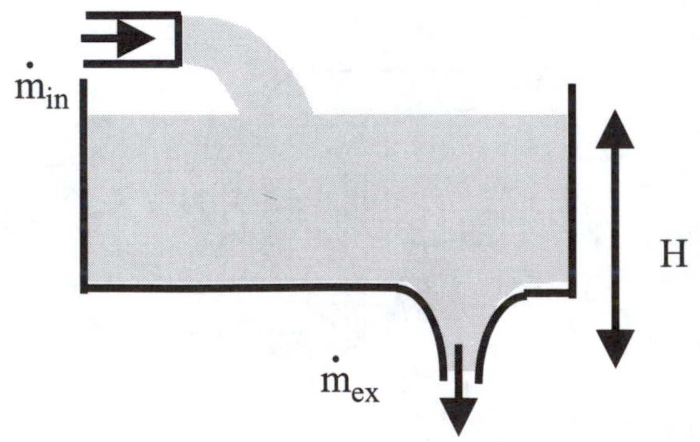

CHAPTER 10
STUDY PROBLEMS

IRREVERSIBILITY AND AVAILABILITY

- **Available energy, reversible work and irreversibility**
- **Availability and second-law efficiency**
- **Exergy balance equation**

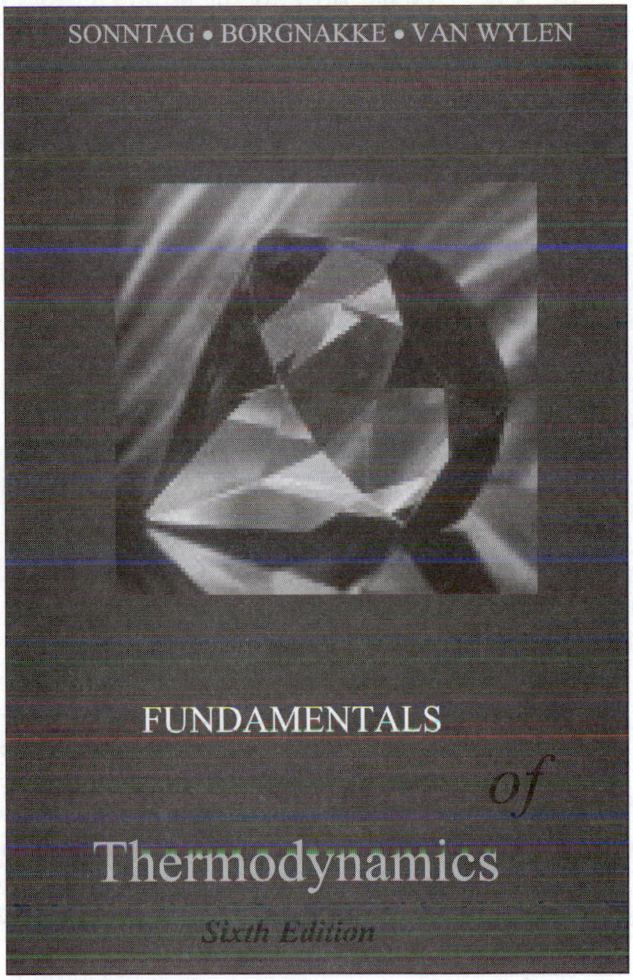

10.1 Availability in a car engine

Find the availability delivered by burning of the fuel at 1800 K in the car for Ex. 7.1. Also find the rate of availability going out with the exhaust flow at 750 K assuming it is half the low temperature heat rejection and that of the other half is heat transfer leaving the engine at 370 K to the ambient at 298 K.

Solution:

From example 7.1 we have:

$$\dot{Q}_H = 333 \text{ kW} \quad \text{at} \quad T_H = 1800 \text{ K}$$

$$\dot{Q}_L = 233 \text{ kW} \quad \text{exhaust flow and heat transfer to air}$$

The rate of fuel availability is therefore the power we could extract from a Carnot heat engine receiving the fuel energy and rejecting heat transfer to the ambient (see Eq. 10.1)

$$\dot{\Phi}_{fuel} = \eta_{HE}\, \dot{Q}_H = \left(1 - \frac{T_o}{T_H}\right) \dot{Q}_H = \left(1 - \frac{298}{1800}\right) 333 \text{ kW} = 278 \text{ kW}$$

The exhaust energy flow is $0.5\, \dot{Q}_L = 117$ kW so here we get

$$\dot{\Phi}_{exh\ flow} = \eta_{HE}\, 0.5\, \dot{Q}_L = \left(1 - \frac{T_o}{T_{exh}}\right) 0.5\, \dot{Q}_L = \left(1 - \frac{298}{750}\right) 117 \text{ kW} = 70 \text{ kW}$$

Now the heat transfer by the coolant to the ambient gives

$$\dot{\Phi}_{exh\ flow} = \eta_{HE}\, 0.5\, \dot{Q}_L = \left(1 - \frac{T_o}{T_{htr}}\right) 0.5\, \dot{Q}_L = \left(1 - \frac{298}{370}\right) 117 \text{ kW} = 23 \text{ kW}$$

Comment: Notice that the availability becomes a lower fraction of the heat transfer as the temperature drops towards the ambient T. Once the heat transfer reaches the ambient it has zero availability left.

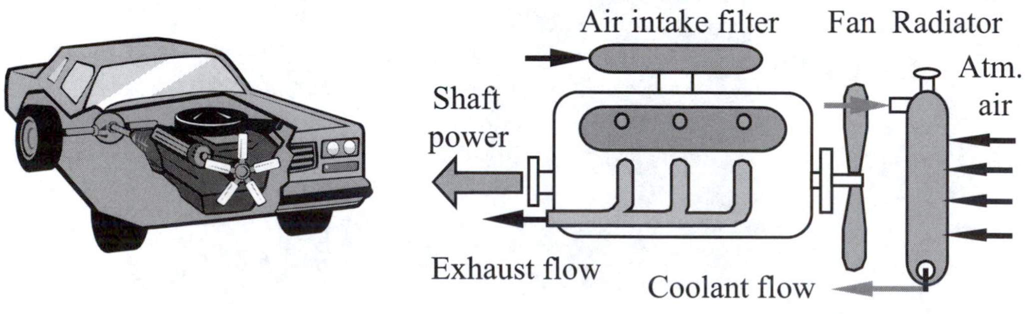

10.2 A heat engine with finite ΔT heat transfers

In a real heat engine we must have finite temperature differences to drive the heat transfer in and out. Assume the heat engine in Ex. 7.3 has an inside cycle high temperature of 500^oC and a low cycle temperature of 350 K and otherwise operates in a Carnot cycle. What are the rates of availability in and out?

Solution:

A Carnot cycle with high temperature $T_a = 500^oC$ and low temperature of $T_b = 350$ K has an efficiency of

$$\eta_{HE} = \left(1 - \frac{T_b}{T_a}\right) = \left(1 - \frac{350}{500 + 273}\right) = 0.547$$

with an input of 1 MW the power output becomes

$$\dot{W} = \eta_{HE}\,\dot{Q}_H = 0.547 \times 1000 \text{ kW} = 547 \text{ kW}$$

which is 100% availability. The rate of availability from the original source is

$$\dot{\Phi}_{Hsource} = \left(1 - \frac{T_o}{T_H}\right)\dot{Q}_H = \left(1 - \frac{300}{550 + 273}\right) 1000 \text{ kW} = 635 \text{ kW}$$

and that into the cycle is

$$\dot{\Phi}_{Hcycle} = \left(1 - \frac{T_o}{T_a}\right)\dot{Q}_H = \left(1 - \frac{300}{500 + 273}\right) 1000 \text{ kW} = 612 \text{ kW}$$

so we lost 23 kW of availability in the transfer. The low temperature heat rejection rate is 1000 – 547 = 453 kW and its availability becomes

$$\dot{\Phi}_{Lcycle} = \left(1 - \frac{T_o}{T_b}\right)\dot{Q}_L = \left(1 - \frac{300}{350}\right) 453 \text{ kW} = 65 \text{ kW}$$

The engine thus rejects 453 kW of energy but only 65 kW of availability and as the energy reaches ambient T there is zero availability (exergy) left.

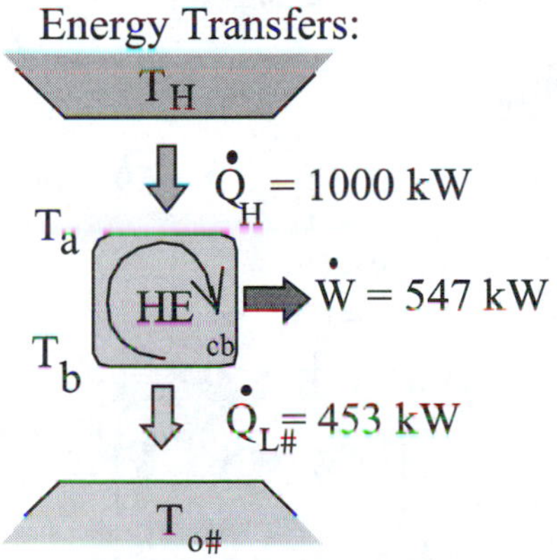

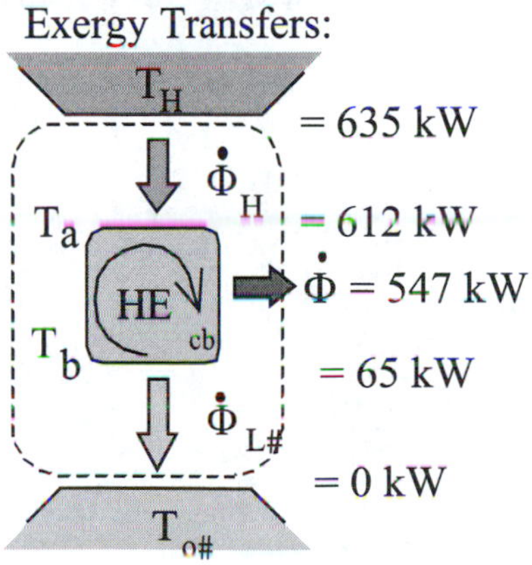

10.3 A steady flow mixing process

A flow of 1 kg/s air at 300 K, 1000 kPa is throttled down to 500 kPa and mixed with another flow of 2 kg/s air at 1000 K, 500 kPa in an open mixing chamber producing a uniform exit flow at 500 kPa. What is the rate of exergy destruction in this process?

Solution:

CV. Throttle (valve) and the mixing chamber.

Continuity Eq.6.9: $\dot{m}_1 + \dot{m}_2 = \dot{m}_3 = 1 + 2 = 3$ kg/s

Energy Eq.6.10: $\dot{m}_1 h_1 + \dot{m}_2 h_2 = \dot{m}_3 h_3$

Entropy Eq.9.7: $\dot{m}_1 s_1 + \dot{m}_2 s_2 + \dot{S}_{gen} = \dot{m}_3 s_3$

Exergy Eq.10.36 $\dot{m}_1 \psi_1 + \dot{m}_2 \psi_2 - \dot{\Phi}_{destruction} = \dot{m}_3 \psi_3$

Solve now (more accurately) using Table A.7.1 and Eq.8.28 for change in s.

$$h_3 = (\dot{m}_1/\dot{m}_3)h_1 + (\dot{m}_2/\dot{m}_3)h_2 = \frac{1}{3} \times 300.47 + \frac{2}{3} \times 1046.22 = 797.637 \text{ kJ/kg}$$

From A.7.1: $T_3 = 777.58$ K $\qquad s^o_{T3} = 7.85396$ kJ/kg K

$$\dot{S}_{gen} = \dot{m}_1(s_3 - s_1) + \dot{m}_2(s_3 - s_2)$$
$$= 1[7.85396 - 6.86926 - 0.287 \ln(500/1000)] + 2(7.85396 - 8.13493)$$
$$= 0.622 \text{ kW/K}$$

The total exergy destruction is

$$\dot{\Phi}_{destruction} = \dot{m}_1 \psi_1 + \dot{m}_2 \psi_2 - \dot{m}_3 \psi_3 = T_o \dot{S}_{gen} = 298.15 \times 0.622 = \mathbf{185 \text{ kW}}$$

We can specify what is destroyed in the valve and the mixing process.

$$\dot{\Phi}_{destruction\ valve} = \dot{m}_1(\psi_1 - \psi_{1a}) = \dot{m}_1[h_1 - h_{1a} - T_o(s_1 - s_{1a})]$$
$$= \dot{m}_1 T_o(s_{1a} - s_1) = 1 \times 298.15\ [-0.287 \ln(\frac{500}{1000})] = \mathbf{59 \text{ kW}}$$

so the balance 185 – 59 = 126 kW is destroyed in the mixing process.

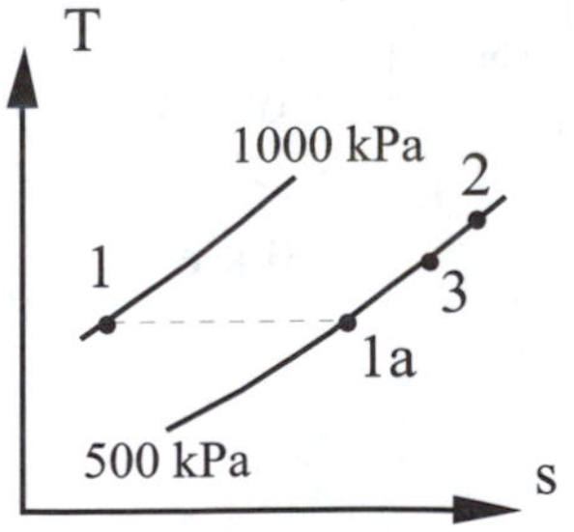

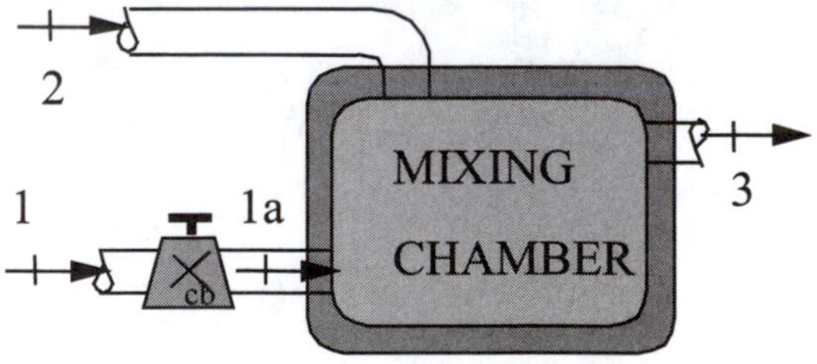

10.4 A feedwater heater

A closed feedwater heater in a powerplant is shown in the figure. It receives steam from a turbine at state 6: 1 MPa, 250°C which gives heat to the feedwater and leaves as saturated liquid at 1 MPa, state 6a. The feedwater, 15 kg/s, comes in at state 2: 5 MPa, 100°C and leaves at state 3: 180°C. We want to find the second law efficiency of this heat exchanger and the rate of exergy destruction.

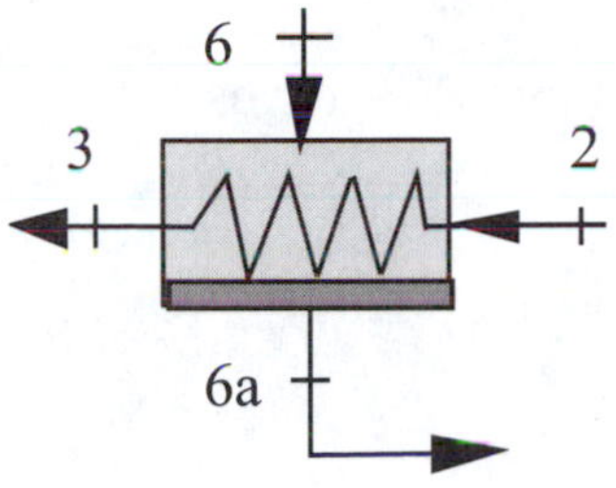

From table B.1	h kJ/kg	s kJ/kgK
B.1.4: 100°C, 5 MPa	$h_2 = 422.71$	$s_2 = 1.303$
B.1.4: 180°C, 5 MPa	$h_3 = 765.24$	$s_3 = 2.1341$
B.1.3: 1 MPa, 250°C	$h_6 = 2942.59$	$s_6 = 6.9246$
B.1.2: 1 MPa, sat. liq.	$h_{6a} = 762.79$	$s_{6a} = 2.1386$

C.V. Feedwater Heater

Energy Eq.: $\dot{m}_2h_2 + \dot{m}_6h_6 = \dot{m}_2h_3 + \dot{m}_6h_{6a}$

Entropy Eq.: $\dot{m}_2s_2 + \dot{m}_6s_6 + \dot{S}_{gen} = \dot{m}_2s_3 + \dot{m}_6s_{6a}$

As all states are known we can solve for the mass flowrate $\dot{m}_6$ from the energy equation

$$\dot{m}_6 = \dot{m}_2 \frac{h_3 - h_2}{h_6 - h_{6a}} = 15 \frac{765.24 - 422.71}{2942.59 - 762.79} = 2.357 \text{ kg/s}$$

And then the entropy generation from the entropy equation

$$\dot{S}_{gen} = \dot{m}_2s_3 + \dot{m}_6s_{6a} - \dot{m}_2s_2 - \dot{m}_6s_6 = \dot{m}_2(s_3 - s_2) + \dot{m}_6(s_{6a} - s_6)$$
$$= 15\ (2.1341 - 1.303) + 2.357(2.1386 - 6.9246)$$
$$= 12.4665 - 11.2806 = 1.1859 \text{ kW/K}$$

The second law efficiency is from Eq.10.30 the ratio of the exergy pick-up $\dot{m}_2(\psi_3 - \psi_2)$ to the exergy delivered by the source $\dot{m}_6\ (\psi_6 - \psi_{6a})$.

$$\dot{\Phi}_{gain} = \dot{m}_2(\psi_3 - \psi_2) = \dot{m}_2[h_3 - h_2 - T_o(s_3 - s_2)]$$
$$= 15\ [765.24 - 422.71 - 298.15(2.1341 - 1.303)]$$
$$= 1421.1 \text{ kW}$$

$$\dot{\Phi}_{source} = \dot{m}_6(\psi_6 - \psi_{6a}) = \dot{m}_6[h_6 - h_{6a} - T_o(s_6 - s_{6a})]$$
$$= 2.357\ [2942.59 - 762.79 - 298.15(6.9246 - 2.1386)]$$
$$= 1774.5 \text{ kW}$$

So now

$$\eta_{II} = \frac{\dot{m}_2(\psi_3 - \psi_2)}{\dot{m}_6\ (\psi_6 - \psi_{6a})} = \frac{1421.1}{1774.5} = \mathbf{0.80}$$

The destruction of exergy term becomes the balance between the two flow terms, Eq.10.36 for steady flow **or** from the entropy generation

$$\dot{\Phi}_{destruction} = T_o\ \dot{S}_{gen} = \dot{m}_6(\psi_6 - \psi_{6a}) - \dot{m}_2(\psi_3 - \psi_2)$$
$$= 298.15 \times 1.1859 = 353.6 \text{ kW}$$
$$= 1774.5 - 1421.1 = \mathbf{353.4\ kW}$$

So the process transferred 80% of the exergy from one flow to the other loosing 353 kW of exergy in the process. It transferred 100% of the energy, $\dot{m}_6(h_6 - h_{6a}) = 5138$ kW from one flow to the other.

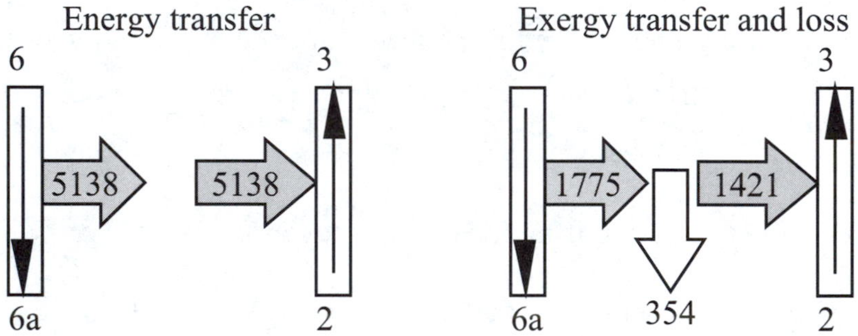

10.5 Stored exergy in a tank

Water at 100 kPa, quality x = 0.4 of mass 0.1 kg is in a rigid insulated tank. An electric heater is turned on for some time after which the water has reached a state of saturated vapor. How much electric work did we add and what is the increase of the water exergy?

Solution:

C.V. Water.

Energy Eq.: $m (u_2 - u_1) = {}_1Q_2 - {}_1W_2 + W_{electric}$

Process Eqs.: ${}_1Q_2 = 0$ (insulated) ${}_1W_2 = 0$ (rigid)

State 1: (P_1, x_1):

$v_1 = v_f + x_1 v_{fg} = 0.001043 + 0.4 \times 1.69296 = 0.67823 \text{ m}^3/\text{kg}$

$u_1 = u_f + x_1 u_{fg} = 417.33 + 0.4 \times 2088.72 = 1252.82 \text{ kJ/kg}$

$s_1 = s_f + x_1 s_{fg} = 1.3025 + 0.4 \times 6.0568 = 3.7252 \text{ kJ/kg K}$

State 2: $x = 1$ and $v_2 = v_1$ =>

$T = 129.5^\circ\text{C}$, $u_2 = u_g = 2539.4 \text{ kJ/kg}$, $s_2 = s_g = 7.0317 \text{ kJ/kg K}$

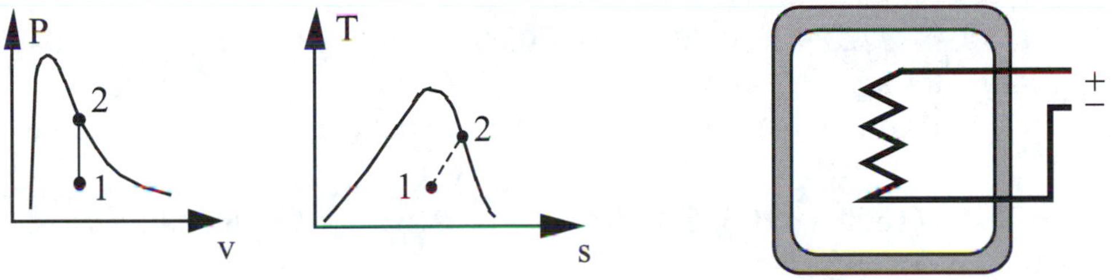

We can now solve for the electric work from the energy equation

$$W_{electric} = m (u_2 - u_1) = 0.1 (2539.4 - 1252.82) = \mathbf{128.66 \text{ kJ}}$$

The water exergy increase is from Eq.10.27

$$\begin{aligned}\phi_2 - \phi_1 &= u_2 + P_o v_2 - T_o s_2 - (u_1 + P_o v_1 - T_o s_1) \\ &= u_2 - u_1 + P_o(v_2 - v_1) - T_o(s_2 - s_1) \\ &= u_2 - u_1 + 0 - T_o(s_2 - s_1)\end{aligned}$$

$$\begin{aligned}\Phi_2 - \Phi_1 &= m(\phi_2 - \phi_1) = m (u_2 - u_1) - T_o m (s_2 - s_1) \\ &= 128.66 - 298.15 \times 0.1 \times (7.0317 - 3.7252) = \mathbf{30.1 \text{ kJ}}\end{aligned}$$

The 128.7 kJ that was added as electric work (100% exergy) gave only 30.1 kJ increase in exergy of the water which means the process destroyed 98.6 kJ of exergy.

10.6 An internally reversible, externally irreversible process

Ammonia in a piston-cylinder that maintains constant P of 1600 kPa with ambient temperature 20°C is now heated to 60°C in an internally reversible process. The external heat source is at a constant 70°C. We want to find the storage, destruction and transfer of exergy in the process.

Solution:

The basic problem is analyzed and solved in study problem 8.4 so let us only look at the exergy related aspects of the problem here and assume an ambient at 20°C.

The heat transfer from the source was found to be

$$ {}_1q_2 = u_2 - u_1 + {}_1w_2 = 1389.3 - 272.89 + 140.6 = 1257 \text{ kJ/kg} $$

which has an exergy of

$$ \phi_{Hsource} = \left(1 - \frac{T_o}{T_H}\right) q_H = \left(1 - \frac{293.15}{70 + 273.15}\right) 1257 \text{ kJ/kg} = \mathbf{183.16\ kJ/kg} $$

The stored exergy is the change in the ammonia exergy as

$$ \begin{aligned} \phi_2 - \phi_1 &= u_2 + P_o v_2 - T_o s_2 - (u_1 + P_o v_1 - T_o s_1) \\ &= u_2 - u_1 + P_o(v_2 - v_1) - T_o(s_2 - s_1) \\ &= 1389.3 - 272.89 + 101(0.08951 - 0.001638) - 293.15\,(5.0472 - 1.0408) \\ &= \mathbf{-49.2\ kJ/kg} \end{aligned} $$

The work out minus the work to the atmosphere is exergy so we have

$$ \begin{aligned} \phi_{out} &= {}_1w_2 - P_o(v_2 - v_1) = (P - P_o)\,(v_2 - v_1) \\ &= (1600 - 101.325)\,(0.08951 - 0.001638) = \mathbf{131.69\ kJ/kg} \end{aligned} $$

The destroyed amount of exergy is

$$ \phi_{destruction} = T_o\ {}_1s_{2\ gen\ tot} = 293.15 \times 0.343 = \mathbf{100.6\ kJ/kg} $$

These terms fit the overall balance as Eq.10.42

$$ \phi_2 - \phi_1 = \phi_{Hsource} - \phi_{out} - \phi_{destruction} $$

$$ -49.2 = 183.16 - 131.69 - 100.6 = -49.13 \quad \text{(OK)} $$

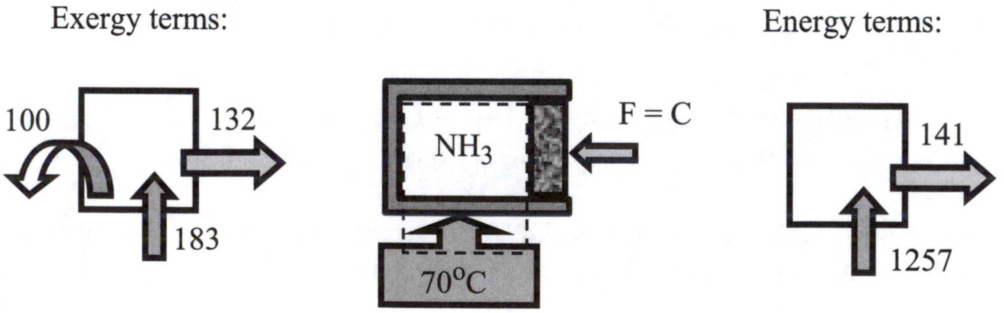

10.7 The air compressor and tank

Consider the air compressor charging a tank with air in Example 9.6, page 312. Find the change in the air exergy.

The initial state of the air in the tank equals the ambient inlet state of the air and for the process we found that $s_2 = s_1$. This also implies $\phi_{in} = \phi_1$

$$\Phi_2 - \Phi_1 = m_1(\phi_2 - \phi_1) + m_{in}(\phi_2 - \phi_{in}) = m_2(\phi_2 - \phi_1)$$

The air exergy increase is from Eq.10.27

$$\begin{aligned}\phi_2 - \phi_1 &= u_2 + P_o v_2 - T_o s_2 - (u_1 + P_o v_1 - T_o s_1)\\ &= u_2 - u_1 + P_o(v_2 - v_1) - T_o(s_2 - s_1)\\ &= u_2 - u_1 + P_o(v_2 - v_1) - 0\end{aligned}$$

So we need the specific volumes

$$v_1 = RT_1/P_1 = 0.287 \times 290.15 / 100 = 0.83273 \text{ m}^3/\text{kg}$$

$$v_2 = RT_2/P_2 = 0.287 \times 555.7 / 1000 = 0.1595 \text{ m}^3/\text{kg}$$

$$\begin{aligned}\Phi_2 - \Phi_1 &= m_2(\phi_2 - \phi_1) = m_2\,[(u_2 - u_1) + P_o(v_2 - v_1)]\\ &= 0.2508\,(401.49 - 207.19 + 100 \times (0.1595 - 0.8327)\\ &= 48.73 - 16.885 = \mathbf{31.85\ kJ}\end{aligned}$$

The increase in air exergy exactly matches the work input to the process. Why is that? Remember the process was assumed to be reversible ($s_{gen} = 0$) and we therefore do not have any exergy destruction (irreversibility).

10.8 A real steam turbine and its 1st and 2nd law efficiencies

A steam turbine receives steam at 2500 kPa, 500°C with an exhaust at 500 kPa, 300°C. We would like to find the first law isentropic efficiency, the specific reversible work, the specific irreversibility and the second law efficiency.

Solution:

C.V. The steam turbine. Steady flow, single flow and we assume adiabatic.

Energy Eq.6.13: $h_i + 0 = h_e + w$

Entropy Eq.9.8: $s_e - s_i = \int_i^e \frac{dq}{T} + s_{gen} = s_{gen}$

For the **actual turbine** we have the inlet and exit states so from steam tables B.1.3

Inlet: $h_i = 3462.04$ kJ/kg, $s_i = 7.3233$ kJ/kg K

Exit: $h_e = 3064.2$ kJ/kg, $s_e = 7.4598$ kJ/kg K

The energy equation gives the actual work as

$$w_{ac} = h_i - h_e = 3462.04 - 3064.2 = 397.84 \text{ kJ/kg}$$

The first law isentropic efficiency is obtained by comparing the actual turbine work to that of a reversible adiabatic (thus isentropic) turbine with the same inlet state and the same exit pressure. This exit state has (P = 500 kPa, s = 7.3233 kJ/kg K)

Exit isentropic: (P, s) Interpolate B.1.3: $T_{e\,s} = 263.9°C$, $h_{e\,s} = 2989.4$ kJ/kg

Now the **isentropic turbine** work is from the energy equation

$$w_s = h_i - h_{e\,s} = 3462.04 - 2989.4 = 472.6 \text{ kJ/kg}$$

and the efficiency becomes

$$\eta_{T\,s} = w_{ac} / w_s = \frac{397.84}{472.6} = \mathbf{0.842}$$

The **reversible work** is from Eq.10.9 with no heat transfer (q = 0) and $T_o = 298$ K

$$\begin{aligned} w^{rev} &= T_o(s_e - s_i) - (h_e - h_i) = \psi_i - \psi_e \\ &= T_o(s_e - s_i) + w_{ac} \\ &= 298\,(7.4598 - 7.3233) + 397.84 \\ &= 40.68 + 397.84 = \mathbf{438.5\ kJ/kg} \end{aligned}$$

The irreversibility from Eq.10.11

$$i = w^{rev} - w_{ac} = T_o(s_e - s_i) = q_o^{rev} = \mathbf{40.68\ kJ/kg}$$

which is also the reversible heat transfer in Eq.10.7. The second law efficiency is the actual work measured relative to the change in availability (which is also the reversible work) as in Eq.10.29

$$\eta_{T\,II} = w_{ac} / w^{rev} = 397.84 / 438.5 = \mathbf{0.907}$$

10.9 Exergy destruction and its location due to heat transfer

Hot combustion gases at 1500 K loses 3000 W to the steel walls at 700 K which in turn delivers the 3000 W to a flow of glycol at 100°C and the glycol has heat transfer to atmospheric air at 20°C. Assume steady state and find the transfer and destruction terms of exergy.

Solution:

The basic problem was solved and examined for entropy in study problem 8.9. We will find the corresponding terms of exergy. The flux of exergy due to the heat transfer is

$$\dot{\Phi} = \left(1 - \frac{T_o}{T}\right)\dot{Q}$$

And the destruction is the difference of these as there are no other terms.

$$\dot{\Phi}_{destruction} = \dot{\Phi}_{in} - \dot{\Phi}_{out} = T_o \; \dot{S}_{gen}$$

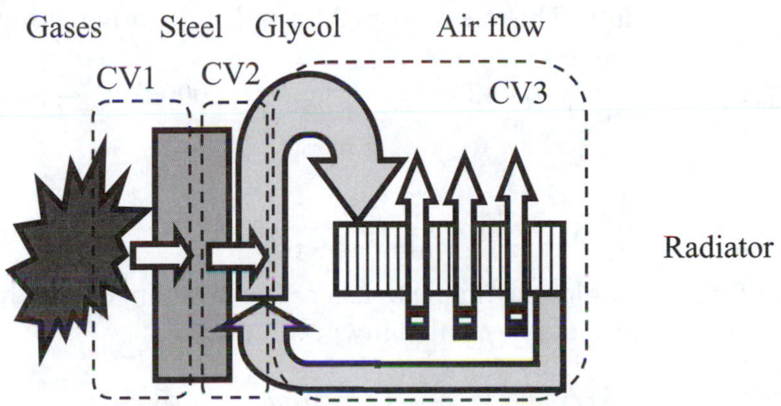

Exergy fluxes and destruction terms.

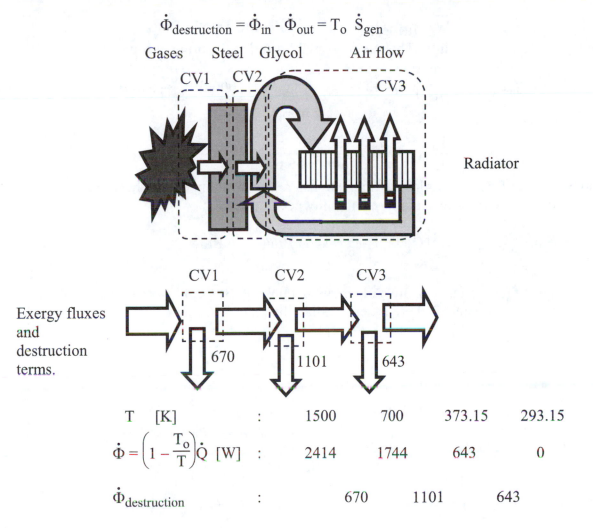

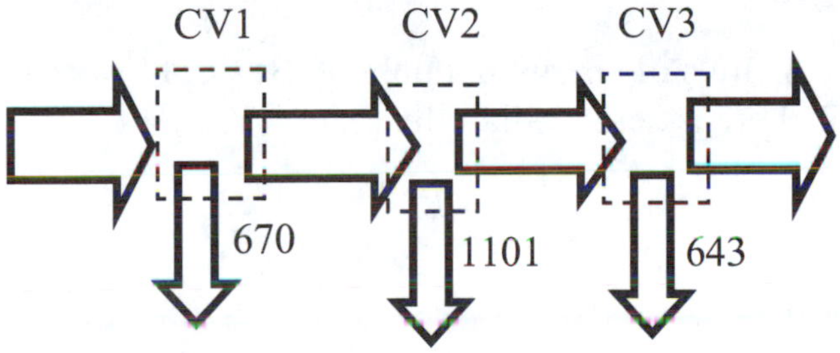

T [K]	:	1500	700	373.15	293.15
$\dot{\Phi} = \left(1 - \frac{T_o}{T}\right)\dot{Q}$ [W]	:	2414	1744	643	0
$\dot{\Phi}_{destruction}$	:	670	1101	643	

10.2E A heat engine with finite ΔT heat transfers

In a real heat engine we must have finite temperature differences to drive the heat transfer in and out. Assume a heat engine as in Ex. 7.3 receiving 1000 Btu/s at 1000 F with an inside cycle high temperature of 900 F and a low cycle temperature of 630 R rejecting heat to an ambient at 540 R and otherwise operates in a Carnot cycle. What are the rates of availability in and out?

Solution:

A Carnot cycle with high temperature T_a = 900 F and low temperature of T_b = 630 R has an efficiency of

$$\eta_{HE} = \left(1 - \frac{T_b}{T_a}\right) = \left(1 - \frac{630}{900 + 459.67}\right) = 0.537$$

with an input of 1 MW the power output becomes

$$\dot{W} = \eta_{HE}\, \dot{Q}_H = 0.537 \times 1000 \text{ Btu/s} = 537 \text{ Btu/s}$$

which is 100% availability. The rate of availability from the original source is

$$\dot{\Phi}_{Hsource} = \left(1 - \frac{T_o}{T_H}\right)\dot{Q}_H = \left(1 - \frac{540}{1000 + 459.67}\right) 1000 \text{ Btu/s} = 630 \text{ Btu/s}$$

and that into the cycle is

$$\dot{\Phi}_{Hcycle} = \left(1 - \frac{T_o}{T_a}\right)\dot{Q}_H = \left(1 - \frac{540}{900 + 459.67}\right) 1000 \text{ Btu/s} = 603 \text{ Btu/s}$$

so we lost 27 Btu/s of availability in the transfer. The low temperature heat rejection rate is 1000 – 537 = 463 Btu/s and its availability becomes

$$\dot{\Phi}_{Lcycle} = \left(1 - \frac{T_o}{T_b}\right)\dot{Q}_L = \left(1 - \frac{540}{630}\right) 463 \text{ Btu/s} = 66 \text{ Btu/s}$$

The engine thus rejects 463 Btu/s of energy but only 66 Btu/s of availability and as the energy reaches ambient T there is zero availability (exergy) left.

Energy Transfers:

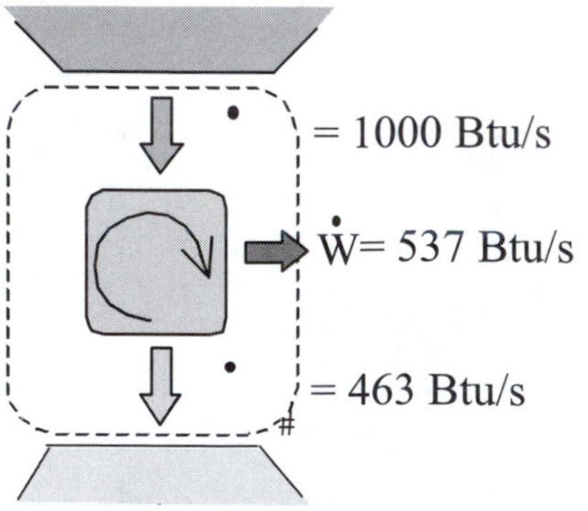

Exergy Transfers:

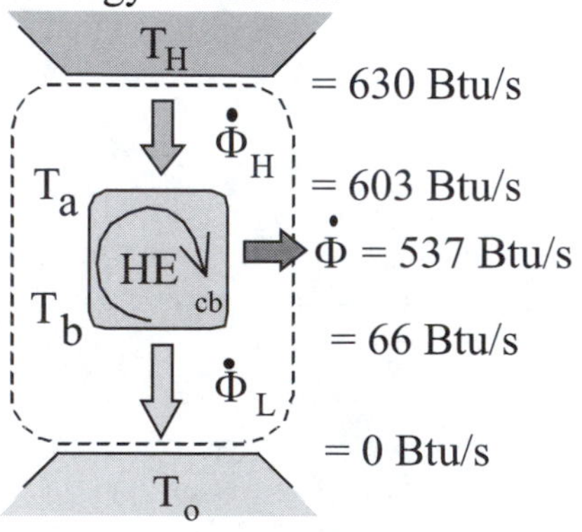

10.6E An internally reversible, externally irreversible process

Ammonia in a piston-cylinder that maintains constant P of 250 psia with ambient temperature 70 F is now heated to 140 F in an internally reversible process. The external heat source is at a constant 160 F. We want to find the storage, destruction and transfer of exergy in the process.

Solution:

The basic problem is analyzed and solved in study problem 8.4 so let us only look at the exergy related aspects of the problem here.

The heat transfer from the source was found to be

$$ {}_1q_2 = u_2 - u_1 + {}_1w_2 = 595.1 - 119.58 + 59.63 = 535.2 \text{ Btu/lbm} $$

which has an exergy of

$$ \phi_{Hsource} = \left(1 - \frac{T_o}{T_H}\right) q_H = \left(1 - \frac{70 + 459.67}{160 + 459.67}\right) 535.2 \text{ Btu/lbm} = \mathbf{77.73 \text{ Btu/lbm}} $$

The stored exergy is the change in the ammonia exergy as

$$ \phi_2 - \phi_1 = u_2 + P_o v_2 - T_o s_2 - (u_1 + P_o v_1 - T_o s_1) $$
$$ = u_2 - u_1 + P_o(v_2 - v_1) - T_o(s_2 - s_1) $$
$$ = 595.1 - 119.58 + 14.7(1.315 - 0.02631)\frac{144}{778} - 529.67\,(1.193 - 0.2529) $$
$$ = \mathbf{-18.9 \text{ Btu/lbm}} $$

The work out minus the work to the atmosphere is exergy so we have

$$ \phi_{out} = {}_1W_2 - P_o(v_2 - v_1) = (P - P_o)\,(v_2 - v_1) $$
$$ = (250 - 14.7)\,(1.315 - 0.02631)\ 144 / 778 = \mathbf{56.1 \text{ Btu/lbm}} $$

The destroyed amount of exergy is

$$ \phi_{destruction} = T_o\ {}_1s_{2\ gen\ tot} = 529.67 \times 0.0764 = \mathbf{40.5 \text{ Btu/lbm}} $$

These terms fit the overall balance as Eq.10.42

$$ \phi_2 - \phi_1 = \phi_{Hsource} - \phi_{out} - \phi_{destruction} $$
$$ -18.9 = 77.73 - 56.1 - 40.5 = -18.87 \quad \text{(OK)} $$

Exergy terms:

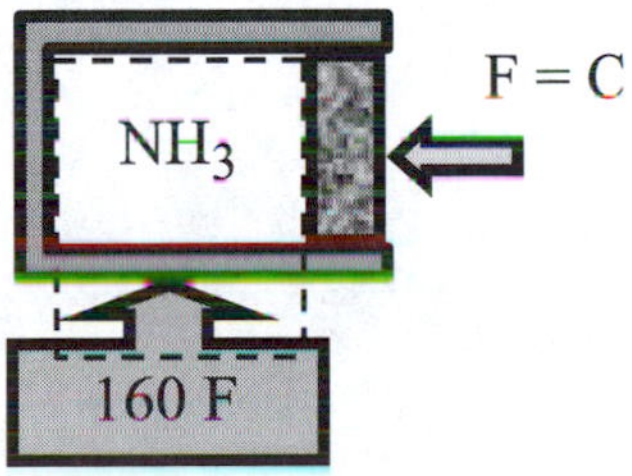

Energy terms:

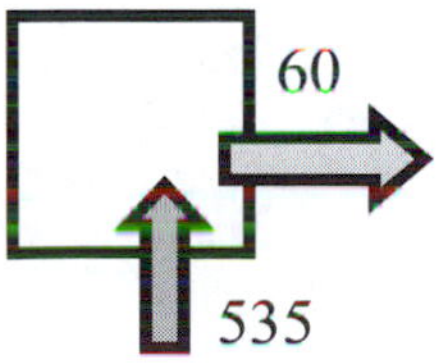

CHAPTER 11
STUDY PROBLEMS

POWER AND REFRIGERATION SYSTEMS

- **The steam power plant, Rankine cycle**
- **The reheat cycle and feedwater heaters**
- **The gas turbine, Brayton cycle, jet engine**
- **Reciprocating engine cycles, Otto, Diesel and Stirling cycles**
- **Vapor-compression refrigeration cycle**
- **Air standard refrigeration cycle and combined cycles**

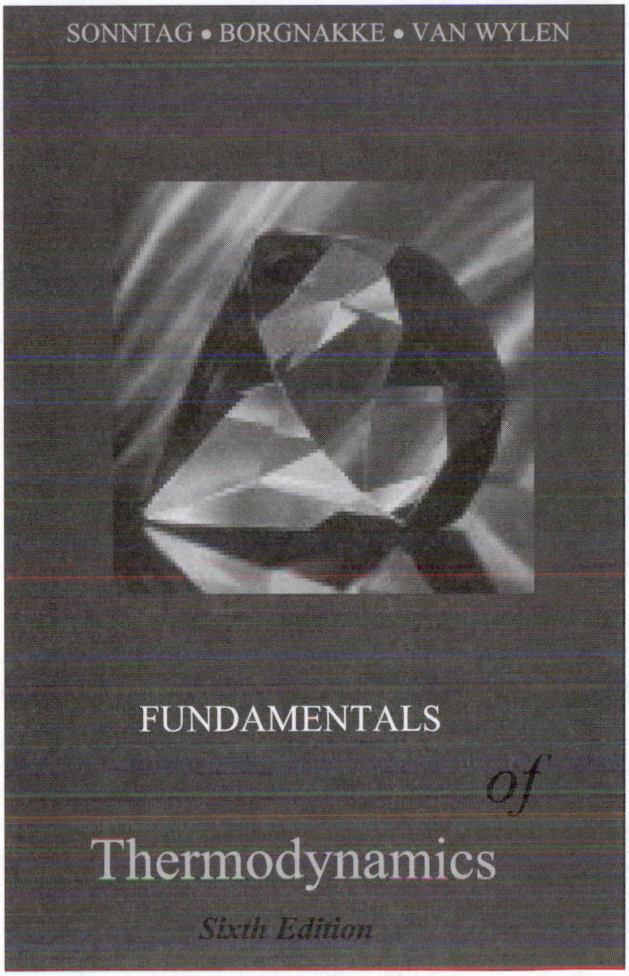

11.1 A simple Rankine cycle

A power plant on the North Pole uses R-22 with a boiler outlet of 3000 kPa, 110 C. The condenser operates at –30 C with the ambient being at –40 C.

a) Find the specific energy transfer in all components.
b) Find the cycle thermal efficiency.

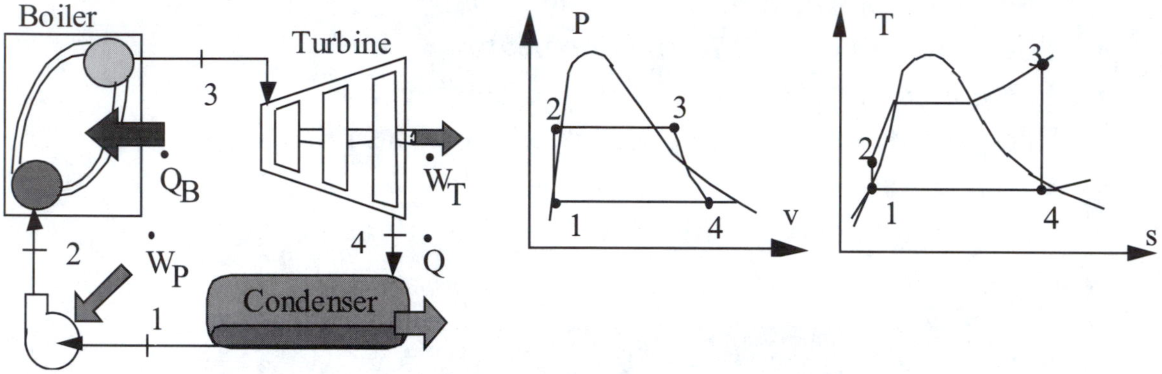

State out of boiler 3 (P, T) Table B.4.2: $h_3 = 306.74$ kJ/kg, $s_3 = 0.9555$ kJ/kg K

State 1 (T_1, $x_1 = 0$) Tbl. B.4.1: $P_1 = 163.5$ kPa, $h_1 = 10.73$ kJ/kg, $v_1 = 0.000725$ m^3/kg

C.V. Pump. Reversible adiabatic: $s_2 = s_1$ implemented from Eq.9.18 with constant v.

$$w_{P1} = v_1 (P_2 - P_1) = 0.000725 \text{ m}^3\text{/kg } (3000 - 163.5) \text{ kPa} = \mathbf{2.06 \text{ kJ/kg}}$$

State out of boiler 3 (P, T) Table B.4.2: $h_3 = 306.74$ kJ/kg, $s_3 = 0.9555$ kJ/kg K

C.V. Turbine reversible, adiabatic: $s_4 = s_3$

$$x_4 = (s_4 - s_f)/s_{fg} = (0.9555 - 0.0449)/0.9335 = 0.97547$$

$$h_4 = h_f + x_4 h_{fg} = 10.73 + 0.\ 97547 \times 227.0 = 232.16 \text{ kJ/kg}$$

$$w_T = h_3 - h_4 = 306.74 - 232.16 = \mathbf{74.58 \text{ kJ/kg}}$$

C.V. Steam generator (boiler)

$$q_H = h_3 - h_2 = 306.74 - 12.79 = \mathbf{293.95 \text{ kJ/kg}}$$

C.V. Condenser

$$q_L = h_4 - h_1 = 232.16 - 10.73 = \mathbf{221.43 \text{ kJ/kg}}$$

Net work

$$w_{net} = w_T - w_{P1} = 74.58 - 2.06 = 72.52 \text{ kJ/kg} \quad (= q_{net} = q_H - q_L)$$

Cycle efficiency: $$\eta_{thermal} = \frac{w_{net}}{q_H} = \frac{72.52}{293.95} = \mathbf{0.247}$$

11.2 A Rankine cycle with reheat

Consider an ideal steam reheat cycle where steam enters the high-pressure turbine at 3.0 MPa, 400°C, and then expands to 0.8 MPa. It is then reheated at the 800 kPa and expands to 10 kPa in the low-pressure turbine. To what temperature should it be reheated to have a minimum quality of 90.116% in the turbine? Calculate the cycle thermal efficiency.

Solution:

C.V. LP Turbine section

State 6: 10 kPa, x = 0.90116 =>

$$h_6 = 191.81 + 0.90116 \times 2392.82 = 2348.12 \text{ kJ/kg}$$

$$s_6 = 0.6492 + 0.90116 \times 7.501 = 7.4088 \text{ kJ/kg K}$$

State 5: 800 kPa, $s_5 = s_6$ =>

$$T_5 = \mathbf{350\ ^oC}, \quad h_5 = 3161.68 \text{ kJ/kg}$$

C.V. HP Turbine section

$P_3 = 3$ MPa, $T_3 = 400°$C => $h_3 = 3230.82$ kJ/kg, $s_3 = 6.9211$ kJ/kg K

$s_4 = s_3$ => $h_4 = 2891.6$ kJ/kg;

C.V. Pump reversible, adiabatic and assume incompressible flow

$w_P = v_1(P_2 - P_1) = 0.00101(3000 - 10) = 3.02$ kJ/kg,

$h_2 = h_1 + w_P = 191.81 + 3.02 = 194.83$ kJ/kg

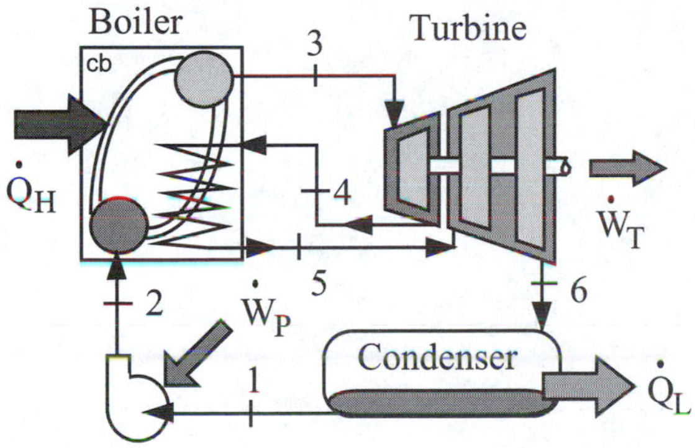

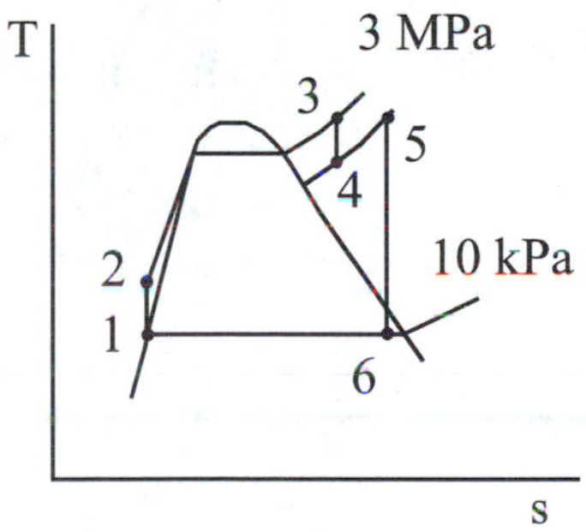

$$w_{T,tot} = h_3 - h_4 + h_5 - h_6 = 3230.82 - 2891.6 + 3161.68 - 2348.12 = 1152.78 \text{ kJ/kg}$$

$$q_{H1} = h_3 - h_2 = 3230.82 - 194.83 = 3036 \text{ kJ/kg}$$

$$q_H = q_{H1} + h_5 - h_4 = 3036 + 3161.68 - 2891.6 = 3306.08 \text{ kJ/kg}$$

$$\eta_{CYCLE} = (w_{T,tot} - w_P)/q_H = (1152.78 - 3.02)/3306.08 = \mathbf{0.348}$$

11.3 An open FWH Rankine cycle

The U of M power plant feeds 25 kg/s steam at 3 MPa, 500°C to the turbine. The condenser temperature is 35°C. Assume the plant operates one open feedwater heater at 200 kPa, where the feedwater leaves as saturated liquid.

a) Find the extraction flowrate.

b) Find the total turbine work.

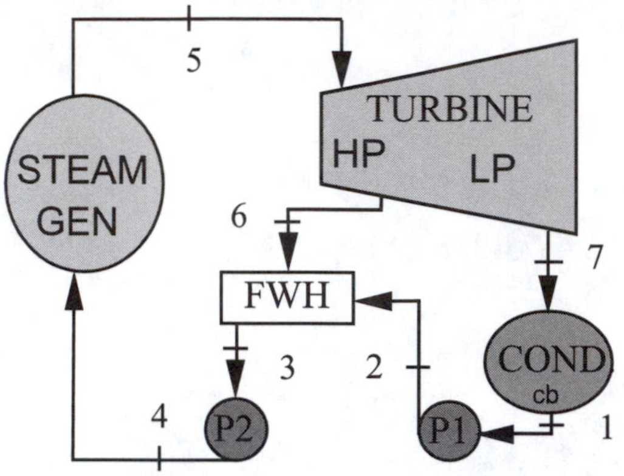

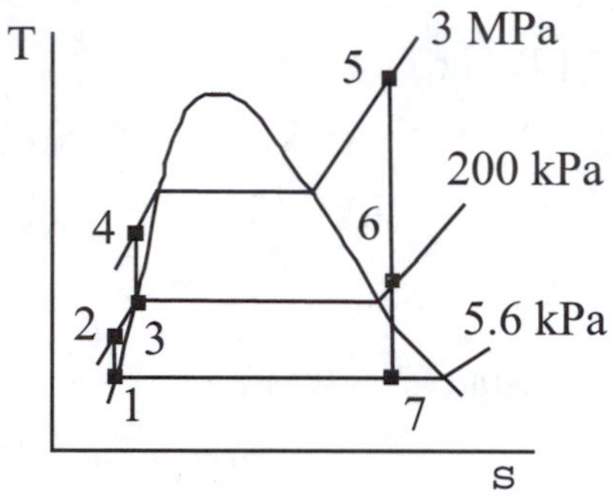

State out of boiler 5: $h_5 = 3456.48$ kJ/kg, $s_5 = 7.2337$ kJ/kg K

C.V. Turbine reversible, adiabatic: $s_7 = s_6 = s_5$

State 6: P_6 , s_6 => $h_6 = 2750$ kJ/kg

State 7: T_7, s_7 => $x_7 = 0.85737$, $h_7 = 2220.3$ kJ/kg

State 3: P_6 , $x_3 = 0$ => $h_3 = 504.68$ kJ/kg

C.V Pump P1

$$w_{P1} = h_2 - h_1 = v_1(P_2 - P_1) = 0.001006(200 - 5.63) = 0.196 \text{ kJ/kg}$$

$$\Rightarrow h_2 = h_1 + w_{P1} = 146.66 + 0.196 = 146.86 \text{ kJ/kg}$$

C.V. Feedwater heater: Call $\dot{m}_6 / \dot{m}_{tot} = x$ (the extraction fraction)

Energy Eq.: $(1 - x)\, h_2 + x\, h_6 = 1\, h_3$

$$x = \frac{h_3 - h_2}{h_6 - h_2} = \frac{504.68 - 146.86}{2750 - 146.86} = \mathbf{0.1375}$$

C.V Turbine

$$\dot{W} = \dot{m}_5 h_5 - x\, \dot{m}_5 h_6 - (1 - x)\, \dot{m}_5 h_7$$
$$= 25 \times 3456.48 - 0.1375 \times 25 \times 2750 - 0.8625 \times 25 \times 2220.3$$
$$= \mathbf{29\ 084\ kW = 29.1\ MW}$$

11.4 A closed FWH Rankine cycle

A closed feedwater heater in a regenerative steam power cycle has an extraction fraction of 17% taken from the turbine at 2 MPa, 350°C. The extraction steam condenses and is throttled to the next lower pressure open feedwater heater. The feedwater comes at 5MPa, 100°C into the heater and leaves at 5MPa. What is the feedwater temperature for the flow going to the boiler?

Solution:

The schematic is from Figure 11.11 has the feedwater from the pump coming at state 2 being heated by the extraction flow coming from the turbine state 6 so the feedwater leaves as hot liquid state 4 and the extraction flow leaves as condensate state 6a.

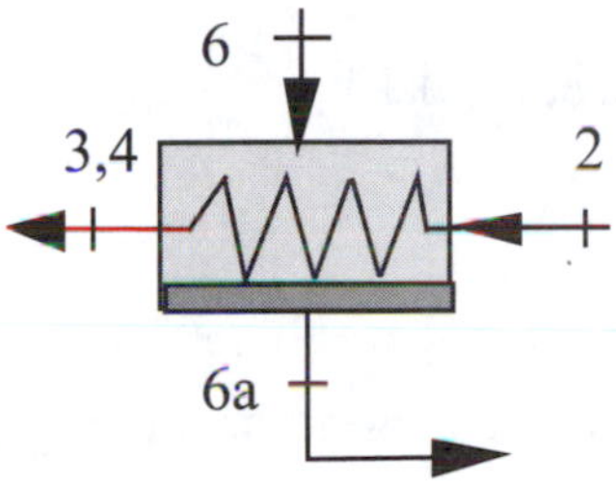

From table B.1	h kJ/kg
B.1.4: 100°C, 5 MPa	$h_2 = 422.71$
B.1.3: 2 MPa, 350°C	$h_6 = 3136.96$
B.1.2: 2 MPa, sat. liq.	$h_{6a} = 908.77$

C.V. Feedwater Heater

Energy Eq.: $\dot{m}_2 h_2 + \dot{m}_6 h_6 = \dot{m}_2 h_4 + \dot{m}_6 h_{6a}$

Since all states except 4 are known we can solve for the enthalpy h_4

$$h_4 = h_2 + \frac{\dot{m}_6}{\dot{m}_2}(h_6 - h_{6a}) = 422.71 + 0.17(3136.96 - 908.77) = 801.5 \text{ kJ/kg}$$

Table B.1.4: (P_4, h_4) => $T_4 =$ **188.5°C**

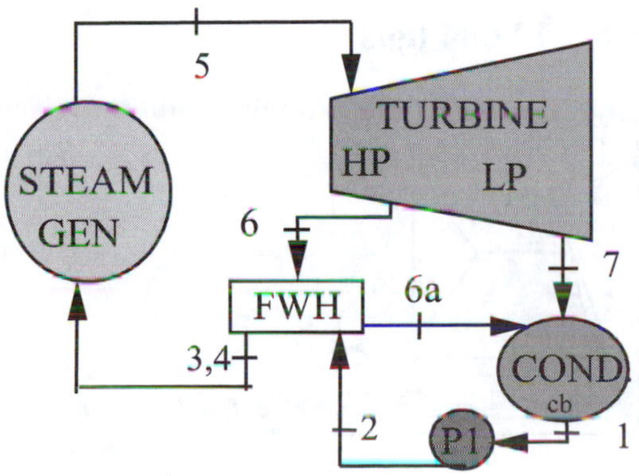

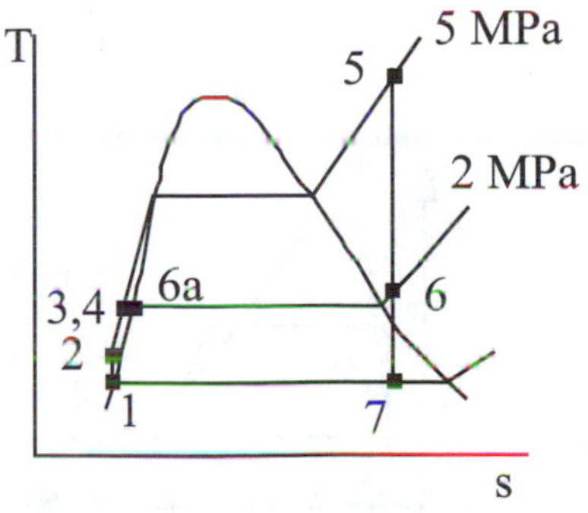

11.5 An actual turbine in a powerplant

A steam power plant has a boiler exit temperature of 900°C and a high pressure of 5 MPa. The turbine produces a power of 20 MW and has an isentropic efficiency of 85% with an exhaust pressure of 10 kPa. We want to find the cycle efficiency and the mass flowrate.

This is a Rankine cycle. Single inlet and exit flows in steady state.

First consider C.V.: Ideal turbine (reversible and adiabatic) inlet:3 exit: 4s

State 3: 900°C, 5 MPa => $h_3 = 4378.8$ kJ/kg ; $s_3 = 7.9593$ kJ/kg K

State 4s: 10 kPa, $s = s_3$ => $x_{4s} = (7.9593\text{-}0.6492)/7.501 = 0.97455$

$$h_{4s} = 191.81 + 0.97455 \times 2392.8 = 2523.7 \text{ kJ/kg}$$

Energy equation: $w_s = h_3 - h_{4s} = 4378.8 - 2523.7 = 1855.1$ kJ/kg

Now consider the actual turbine with exit state 4ac

Energy equation: $w_{ac} = 0.85 \times w_s = 1576.8 = h_3 - h_{4ac}$

$$\Rightarrow h_{4ac} = 2802 \text{ kJ/kg}, \quad T_{4ac} = 160^\circ\text{C}, \quad s_{4ac} = 8.7306 \text{ kJ/kg K}$$

The boiler control volume requires that state 2 is known so we look at pump:

Pump: $w_p = h_2 - h_1 = v_1(P_2 - P_1) = 0.00101(5000 - 10) = 5.04$ kJ/kg

$$h_2 = h_1 + w_p = 191.81 + 5.04 = 196.85 \text{ kJ/kg}$$

Boiler: $q_H = h_3 - h_2 = 4378.8 - 196.85 = 4182$ kJ/kg

Net work = $w_{ac} - w_p = 1576.8 - 5.04 = 1571.8$ kJ/kg

Efficiency = $w_{net}/q_H = 1571.8/4182 =$ **0.376**

The total flow rate is from the total power as:

$$\dot{m}\, w_{ac} = \dot{W}_{ac} \quad \Rightarrow \dot{m} = 20\,000/1576.8 = \mathbf{12.684\ kg/s}$$

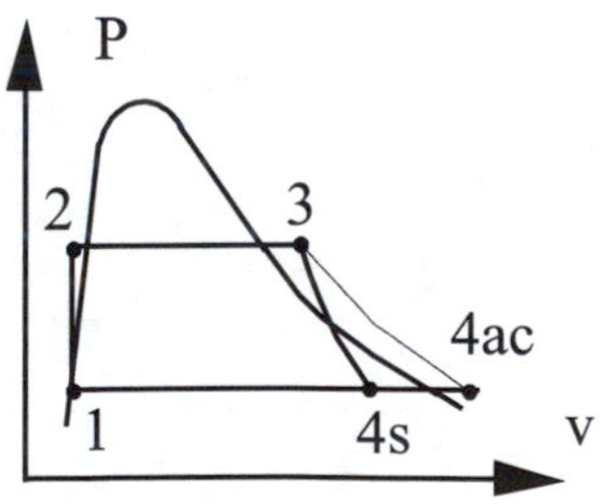

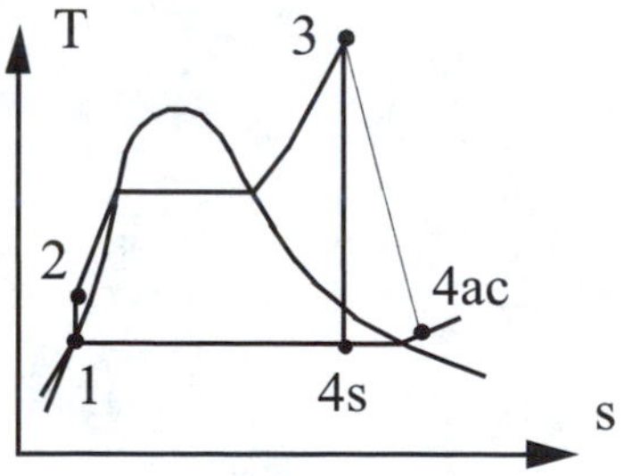

11.6 An actual turbine in a powerplant and the destruction of availability

Let us look at the steam power plant in Study problem 11.5.

a) Find the rate of availability increase of the water in the boiler.

b) Find the rate of availability destruction in the turbine

This is a Rankine cycle. Single inlet and exit flows in steady state.

Now consider the actual turbine with exit state 4ac, given the isentropic work

Energy equation: $w_{ac} = 0.85 \times w_s = 1576.8 \text{ kJ/kg} = h_3 - h_{4ac}$

$$=> h_{4ac} = 2802 \text{ kJ/kg}, \quad T_{4ac} = 160^{o}\text{C}, \quad s_{4ac} = 8.7306 \text{ kJ/kg K}$$

The boiler control volume requires that state 2 is known so we look at pump:

$$\text{Pump:} \quad w_p = h_2 - h_1 = v_1(P_2 - P_1) = 0.00101(5000 - 10) = 5.04 \text{ kJ/kg}$$

$$h_2 = h_1 + w_p = 191.81 + 5.04 = 196.85 \text{ kJ/kg}$$

The availability increase in the boiler is

$$\dot{m}(\psi_3 - \psi_2) = \dot{m}[h_3 - h_2 - T_o(s_3 - s_2)]$$

$$= 12.684 \,[4182 - 298.15\,(7.9593 - 0.6492)]$$

$$= 12.684 \times 2002.5 = 25.4 \text{ MW}$$

The destruction of availability in the turbine is proportional to the entropy generation:

$$s_{gen} = s_{4ac} - s_3 = 8.7306 - 7.9593 = 0.7713 \text{ kJ/kg K}$$

$$\dot{I} = \dot{m}\, T_o\, s_{gen} = 12.684 \times 298.15 \times 0.7713 = \mathbf{2917\ kW}$$

11.7 A gasturbine

A Brayton cycle has an inlet is at 280 K, 80 kPa and the compression ratio is 15:1 with the combustion process adding 1000 kJ/kg. What is the maximum temperature in the cycle? Find also the net specific work and cycle efficiency. Use cold air properties to solve.

Solution:

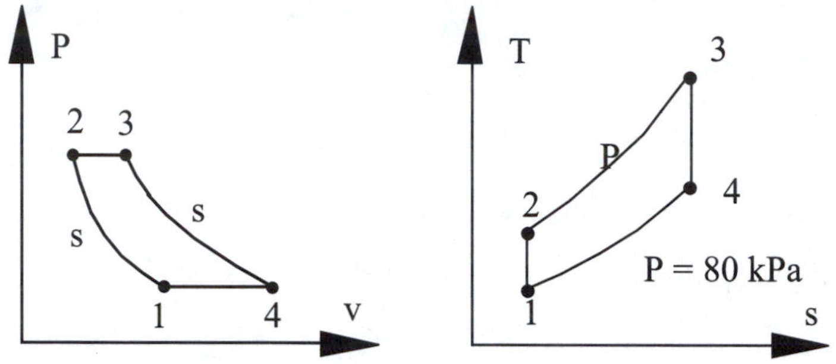

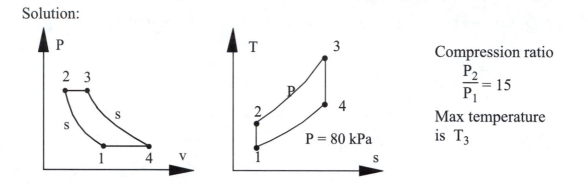

Compression ratio $\frac{P_2}{P_1} = 15$

Max temperature is T_3

Compression process: $h_2 = h_1 + w_C$ and $s_2 = s_1$

$$T_2 = T_1 (P_2 / P_1)^{(k-1)/k} = 280 \times 15^{0.2857} = 607 \text{ K}$$

$$w_C = h_2 - h_1 \approx C_p(T_2 - T_1) = 1.004(607 - 280) = 328.3 \text{ kJ/kg}$$

Combustion: $h_3 = h_2 + q_H$; ${}_2w_3 = 0$

$$T_{max} = T_3 = T_2 + q_H/C_p = 607 + 1000/1.004 = \mathbf{1603\ K}$$

The expansion process: $w_T = h_3 - h_4$ and $s_4 = s_3$

$$T_4 = T_3 (P_4 / P_3)^{(k-1)/k} = 1603 \times (1/15)^{0.2857} = 739.5 \text{ K}$$

$$w_T = h_3 - h_4 \approx C_p(T_3 - T_4) = 1.004(1603 - 739.5) = 866.9 \text{ kJ/kg}$$

The net work and cycle efficiency becomes

$$w_{net} = w_T - w_C = 866.9 - 328.3 = \mathbf{538.6\ kJ/kg}$$

$$\eta = w_{net} / q_H = 538.6 / 1000 = \mathbf{0.54}$$

11.8

Repeat the previous problem but use the air Tables A.7 to solve the problem instead of the cold air properties.

Solution:

From Table A.7.1

$$T_1 = 280 \text{ K}; \quad h_1 = 280.39 \text{ kJ/kg}, \quad s^o_{T1} = 6.79998 \text{ kJ/kg K}$$

Reversible adiabatic compression leads to constant s, from Eq.8.28:

$$s^o_{T2} = s^o_{T1} + R \ln (P_2 / P_1) = 6.79998 + 0.287 \ln(15) = 7.57719 \text{ kJ/kg K}$$

From A.7.1: $h_2 = 607.81$ kJ/kg, $T_2 = 600.5$ K

The constant pressure combustion process (${}_2w_3 = 0$):

$$h_3 = h_2 + q_H = 607.81 + 1000 = 1607.81 \text{ kJ/kg}$$

From A.7.1: $T_3 = \mathbf{1476.8\ K}$; $s^o_{T3} = 8.59309$ kJ/kg K

Reversible adiabatic expansion leads to constant s, from Eq.8.28

$$s^o_{T4} = s^o_{T3} + R \ln(P_4 / P_3) = 8.59309 + 0.287 \ln(1 / 15) = 7.81588 \text{ kJ/kgK}$$

From Table A.7.1 by linear interpolation $T_4 \approx 750.9$ K, $h_4 = 768.58$ kJ/kg

$$w_T = h_3 - h_4 = 1607.81 - 768.58 = 839.2 \text{ kJ/kg}$$

$$w_C = h_2 - h_1 = 607.81 - 280.39 = 327.4 \text{ kJ/kg}$$

$$w_{net} = w_T - w_C = 839.2 - 327.4 = \mathbf{511.8\ kJ/kg}$$

$$\eta = w_{net} / q_H = 511.8 / 1000 = \mathbf{0.51}$$

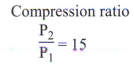

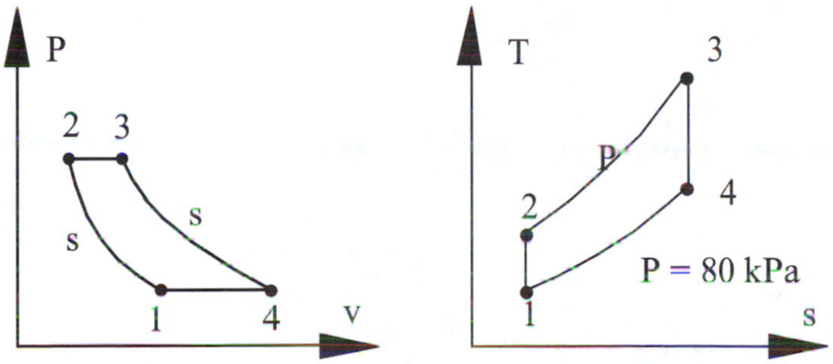

Compression ratio

$$\frac{P_2}{P_1} = 15$$

Max temperature is T_3

11.9 An Otto cycle

A gasoline engine has air at 95 kPa, 290 K before compression. The compression ratio is 9 and for material considerations it is desired to have a maximum temperature of 2000 K. Find the maximum cycle pressure, the specific energy released by combustion and the mean effective pressure using cold air properties.

Solution:

Compression 1 to 2: $s_2 = s_1 \Rightarrow$ From Eq.8.33 and Eq.8.34

$$T_2 = T_1 (v_1/v_2)^{k-1} = 290 \times 9^{0.4} = 698.4 \text{ K}$$

$$P_2 = P_1 \times (v_1/v_2)^{k} = 95 \times 9^{1.4} = 2059 \text{ kPa}$$

Combustion 2 to 3 at constant volume: $u_3 = u_2 + q_H$

$$q_H = u_3 - u_2 = C_v(T_3 - T_2) = 0.717(2000 - 698.4) = \mathbf{933\ kJ/kg}$$

The highest cycle pressure is at state 3

$$P_3 = P_2 \times (T_3/T_2) = 2059\ (2000 / 698.4) = \mathbf{5896\ kPa}$$

To get the mean effective pressure we need the net work, so from Eq.11.18

$$\eta = w_{net} / q_H = 1 - r_v^{(1-k)} = 1 - 9^{-0.4} = 0.585$$

$$w_{net} = \eta \times q_H = 0.585\ 933 = 545.8 \text{ kJ/kg}$$

The displacement is from state 1

$$v_1 = RT_1/P_1 = 0.287 \times 290 / 95 = 0.8761 \text{ m}^3/\text{kg}$$

$$v_2 = v_1 / r_v = 0.8761 / 9 = 0.09734 \text{ m}^3/\text{kg}$$

The mean effective pressure, Eq.11.15, becomes

$$P_{meff} = w_{net} / (v_1 - v_2) = 545.8 / (0.8761 - 0.09734) = \mathbf{700.9\ kPa}$$

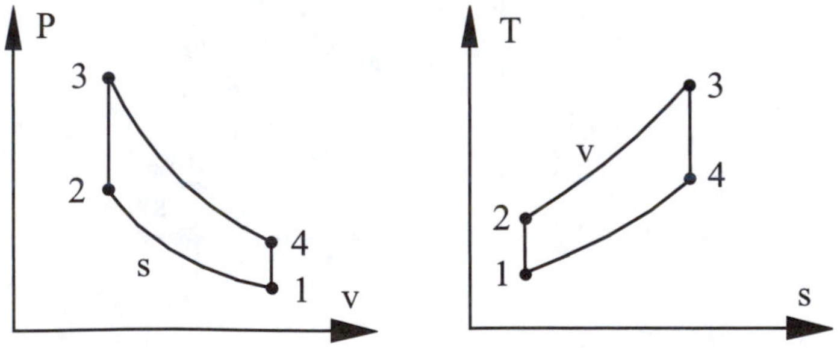

11.10 An actual heat pump

A farmer wants to keep a chicken hatchery at a constant 30°C while the room looses 10 kW to the colder ambient at 10°C. He buys a heat pump operating with ammonia having a high pressure of 1400 kPa and a low temperature of 5°C in the cycle. Find the size of the electric motor that drives the unit. What is the entropy generation rate inside the unit and what is the total entropy generation rate for the whole setup.

Solution:

Nothing was stated about the cycle so we will assume a standard refrigeration cycle with an ideal compressor.

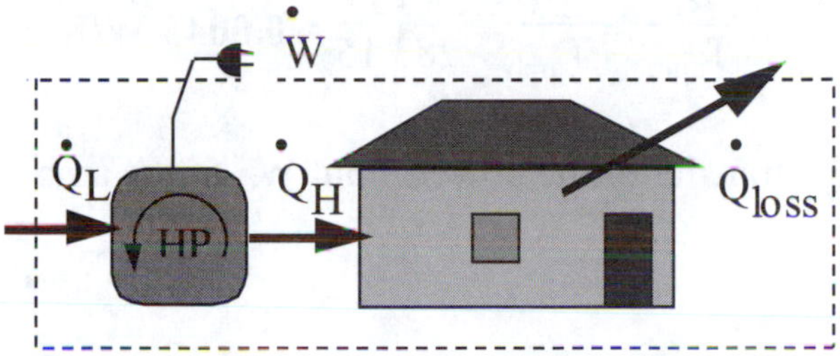

Ideal refrigeration cycle

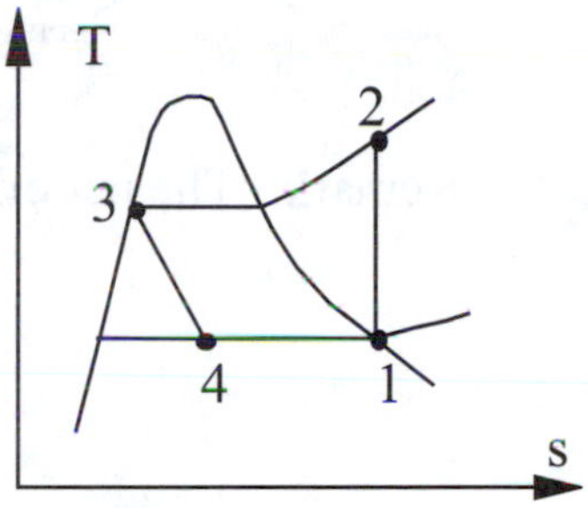

State 1: Inlet to compressor, sat. vapor 5°C, from Table B.2.1

$$h_1 = 1447.3 \text{ kJ/kg}, \quad s_1 = 5.2666 \text{ kJ/kg K}$$

C.V. Ideal Compressor: $w_C = h_2 - h_1, \quad s_2 = s_1$

State 2: 1400 kPa, $s = 5.2666$ kJ/kg K; $T_2 = 75.8°C$, $h_2 = 1587.39$ kJ/kg

$$w_C = h_2 - h_1 = 140.09 \text{ kJ/kg}$$

State 3: Exit condenser, sat. liquid 1400 kPa,

$$h_3 = 353.0 \text{ kJ/kg}, \quad s_3 = 1.2989 \text{ kJ/kg K}$$

State 4: Exit valve, $h_4 = h_3$

$$=> \quad x_4 = (h_4 - h_f)/h_{fg} = \frac{353.0 - 203.58}{1243.7} = 0.12014$$

$$s_4 = s_f + x_4 \, s_{fg} = 0.7951 + 0.12014 \times 4.4715 = 1.3323 \text{ kJ/kg K}$$

C.V. Evaporator: $q_L = h_1 - h_4 = h_1 - h_3 = 1094.3$ kJ/kg

C.V. Condenser: $q_H = h_2 - h_3 = 1234.39$ kJ/kg

$$\dot{Q}_H = \dot{Q}_{loss} = 10 \text{ kW} \quad => \quad \dot{m} = \dot{Q}_H/q_H = \frac{10}{1234.39} = 0.008101 \text{ kg/s}$$

The required work is then

$$\dot{W}_C = \dot{m}\, w_C = 0.008101 \times 140.09 = \mathbf{1.135\ kW}$$

Entropy generation is in the expansion process (throttle or valve)

$$\dot{S}_{gen} = \dot{m}(s_4 - s_3) = 0.008101(1.3323 - 1.2989) = \mathbf{0.00027\ kW/K}$$

Take control volume total all the way out to the ambient 10°C, steady state:

Energy Eq.: $\quad 0 = \dot{W}_C + \dot{Q}_L - \dot{Q}_H$

Entropy Eq.: $\quad 0 = \dfrac{\dot{Q}_L}{T_{amb}} - \dfrac{\dot{Q}_H}{T_{amb}} + \dot{S}_{gen\ tot}$

$$\dot{S}_{gen\ tot} = \frac{\dot{Q}_H}{T_{amb}} - \frac{\dot{Q}_L}{T_{amb}} = \frac{\dot{W}_C}{T_{amb}} = \frac{1.135}{283.15} = \mathbf{0.004\ kW/K}$$

Remark: The net effect is the transfer of the power input $\dot{W}_C$ to the ambient.

11.2E A Rankine cycle with reheat

Consider an ideal steam reheat cycle where steam enters the high-pressure turbine at 600 psia, 700 F, and then expands to 150 psia. It is then reheated at the 150 psia and expands to the pressure in the condenser which is kept at 130 F. To what temperature should it be reheated to have a minimum quality of 88.44% in the turbine? Calculate the cycle thermal efficiency.

Solution:

C.V. LP Turbine section

State 6: 130 F, x = 0.90116 =>

$$h_6 = 97.97 + 0.8844 \times 1019.78 = 999.86 \text{ Btu/lbm}$$

$$s_6 = 0.1817 + 0.8844 \times 1.7292 = 1.711 \text{ Btu/lbm R}$$

State 5: 150 psia, $s_5 = s_6$ =>

$T_5 =$ **600 F**, $h_5 = 1325.69$ Btu/lbm

C.V. HP Turbine section

$P_3 = 600$ psia, $T_3 = 700$ F => $h_3 = 1350.62$ Btu/lbm, $s_3 = 1.5871$ Btu/lbm R

$s_4 = s_3$ => $h_4 = 1208.93$ Btu/lbm;

C.V. Pump reversible, adiabatic and assume incompressible flow

$$w_P = v_1(P_2 - P_1) = 0.01625 \times (600 - 2.225) \times 144 / 778 = 1.8 \text{ Btu/lbm},$$

$$h_2 = h_1 + w_P = 97.97 + 1.8 = 99.77 \text{ Btu/lbm}$$

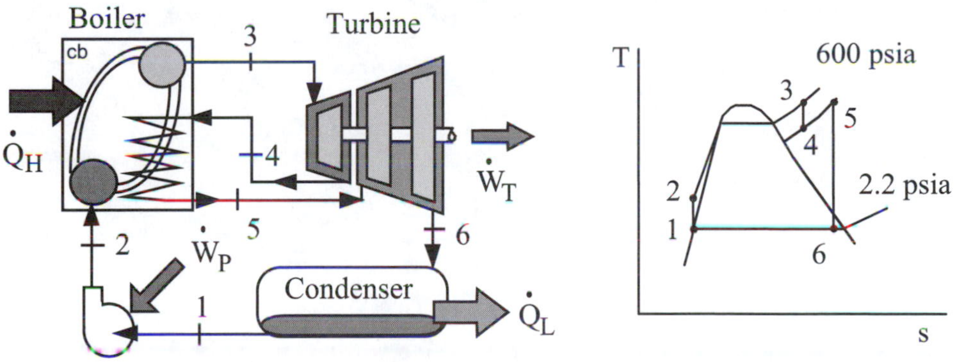

$$w_{T,tot} = h_3 - h_4 + h_5 - h_6 = 1350.62 - 1208.93 + 1325.69 - 999.86$$

$$= 467.52 \text{ Btu/lbm}$$

$$q_{H1} = h_3 - h_2 = 1350.62 - 99.77 = 1250.85 \text{ Btu/lbm}$$

$$q_H = q_{H1} + h_5 - h_4 = 1250.85 + 1325.69 - 1208.93 = \text{ Btu/lbm}$$

$$\eta_{CYCLE} = (w_{T,tot} - w_P)/q_H = (467.52 - 1.8)/1367.6 = \mathbf{0.34}$$

CHAPTER 12
STUDY PROBLEMS

MIXTURES

- **Mixtures of ideal gases**
- **Moist air**
- **The psychrometric chart**

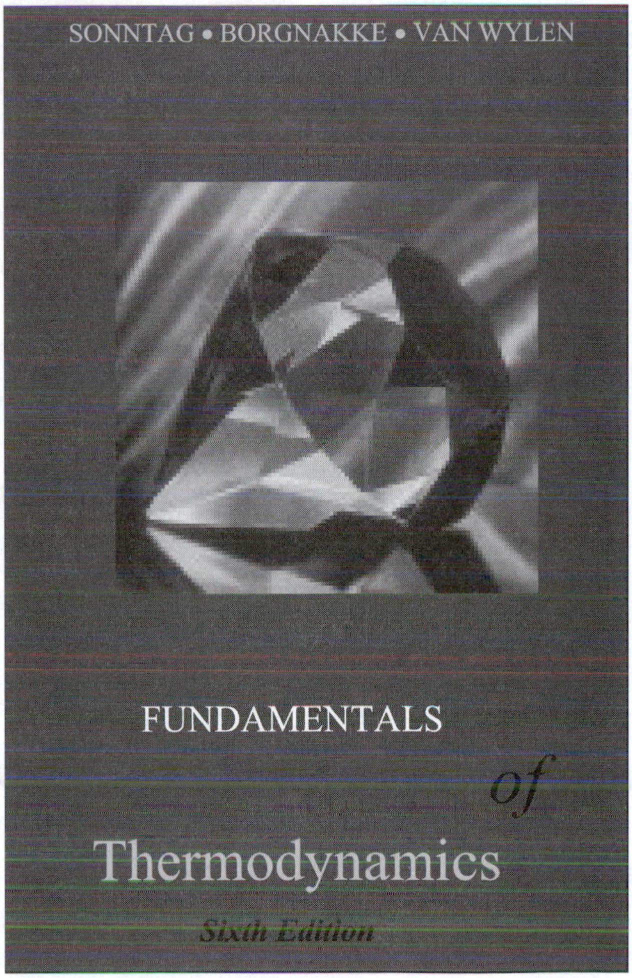

12.1 Mixing in a rigid tank

A mixture of carbon dioxide and oxygen is made by having a rigid tank containing two sections separated by a membrane. In A we have 1 kg CO_2 at 1000 K and in B we have 2 kg O_2 at 300 K both at 200 kPa. The membrane is broken and the mixture comes to a uniform state without any external heat transfer. What is the final temperature and what are the final partial pressures.

Solution:

C.V. The whole tank.

Energy Eq.5.11: $U_2 - U_1 = 0 = m_{CO_2}C_{V0}(T_2 - T_{A1}) + m_{O_2}C_{V0}(T_2 - T_{B1})$

$$V_A = m_{CO_2}RT_{A1}/P = 1 \times 0.1889 \times 1000/\ 200 = 0.9445 \text{ m}^3$$

$$V_B = m_{O_2}RT_{B1}/P = 2 \times 0.2598 \times 300 / 200 = 0.7794 \text{ m}^3$$

Continuity Eq.: $m_2 = m_{CO_2} + m_{O_2} = 3$ kg

Energy Eq.: $1 \times 0.653\ (T_2 - 1000)\ + 2 \times 0.662\ (T_2 - 300) = 0$

Solving, $T_2 = \mathbf{531.2\ K}$

$$R_{mix} = \Sigma\ c_iR_i = \frac{1}{3} \times 0.1889 + \frac{2}{3} \times 0.2598 = 0.2362 \text{ kJ/kg K}$$

$$P_2 = m_2RT_2/(V_A+V_B) = 3 \times 0.2362 \times 531.2 / 1.5588 = \mathbf{241.5\ kPa}$$

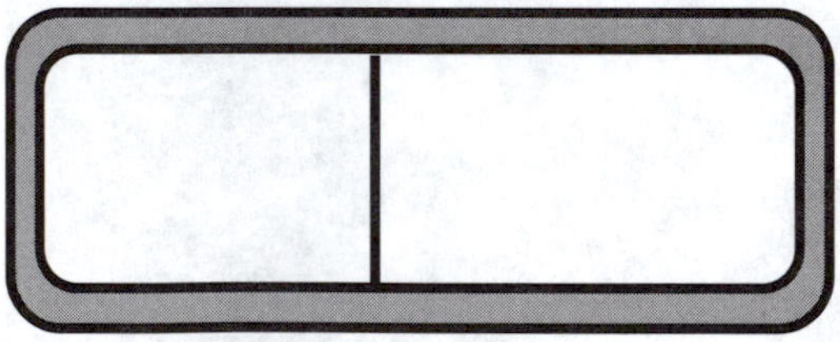

Comment: We could also have used values of u from Table A.8 to write

$$U_2 = (mu)_{CO_2} + (mu)_{O_2} = U_1 = 1 \times 782.75 + 2 \times 195.2 = 1173.15 \text{ kJ}$$

But now it becomes trial and error to find the single T_2. Guess a T_2 and look in A.8

$U_{2 \text{ at } 550 \text{ K}} = 349.12 + 2 \times 367.7 = 1084.5$ kJ; This is too low

$U_{2 \text{ at } 600 \text{ K}} = 392.72 + 2 \times 404.46 = 1201.6$ kJ This a little too high

Interpolate to get: $T_2 = 550 + 50\ (1173.15 - 1084.5)/(1201.6 - 1084.5) = 588$ K

Notice here we got a T_2 that is 10% higher than before so we should have used a higher average value for the heat capacity for CO_2.

12.2 Mixing in a flow setup

A mixture of carbon dioxide and oxygen is made by having a flow of 1 kmol/s CO_2 at 1000 K, 200 kPa mix with a flow of 2 kmol/s O_2 at 300 K, 200 kPa. The exit mixture has a uniform state at 200 kPa total pressure and there is no external heat transfer and no significant kinetic energy. What is the exit temperature and what are the partial pressures in the exit flow?

Solution:

CV mixing chamber, steady flow. The inlet ratio is $2\,\dot{n}_{CO_2} = \dot{n}_{O_2}$ and assume no external heat transfer, no work involved.

Continuity: $\dot{n}_{CO_2} + \dot{n}_{O_2} = \dot{n}_{ex} = 3$ kmol/s; => $y_{O_2} = 2/3$ and $y_{CO_2} = 1/3$

Energy Eq.: $\dot{n}_{CO_2}(2\bar{h}_{O_2} + \bar{h}_{CO_2}) = 3\,\dot{n}_{CO_2}\,\bar{h}_{mix\ ex}$

Take 300 K as reference and write $\bar{h} = \bar{h}_{300} + \bar{C}_{Pmix}(T - 300)$.

$$2\bar{C}_{P\,O_2}(T_{i\,O_2} - 300) + \bar{C}_{P\,CO_2}(T_{i\,CO_2} - 300) = 3\bar{C}_{P\,mix}(T_{mix\ ex} - 300)$$

Find the specific heats in Table A.5 to get

$$\bar{C}_{P\,mix} = \sum y_i\bar{C}_{P\,i} = (0.842 \times 44.01 + 2 \times 0.922 \times 31.999)/3$$

$$= (37.056 + 2 \times 29.503)/3 = 32.02 \text{ kJ/kmol K}$$

The energy equation becomes (notice terms with the 300 K reference drops out)

$$3\bar{C}_{P\,mix}T_{mix\ ex} = 2\bar{C}_{P\,O_2}T_{i\,O_2} + \bar{C}_{P\,CO_2}T_{i\,CO_2} = 54\,757.8 \text{ kJ/kmol}$$

$$T_{mix\ ex} = \mathbf{570\ K}$$

Partial pressures are total pressure times molefraction

$$P_{ex\,O_2} = 2P_{tot}/3 = 133.33 \text{ kPa};\quad P_{ex\,CO_2} = P_{tot}/3 = 66.667 \text{ kPa}$$

$$\dot{S}_{gen} = \dot{n}_{ex}\bar{s}_{ex} - (\dot{n}\bar{s})_{iCO_2} - (\dot{n}\bar{s})_{iO_2} = 2\dot{n}_{CO_2}(\bar{s}_e - \bar{s}_i)_{O_2} + \dot{n}_{CO_2}(\bar{s}_e - \bar{s}_i)_{CO_2}$$

$$\dot{S}_{gen}/3\dot{n}_{CO_2} = [2\bar{C}_{PO_2}\ln\frac{T_{ex}}{T_{iO_2}} - 2\bar{R}\ln y_{O_2} + \bar{C}_{PCO_2}\ln\frac{T_{ex}}{T_{iCO_2}} - \bar{R}\ln y_{CO_2}]/3$$

$$= [\,2(29.503\ln\frac{570}{300} - 8.3145\ln\frac{2}{3}) + 37.056\ln\frac{570}{1000} - 8.3145\ln\frac{1}{3}\,]/3$$

$$= [2(18.9366 + 3.3712) - 20.8299 + 9.1344]/3 = \mathbf{10.97\ kJ/kmol\ mix\ K}$$

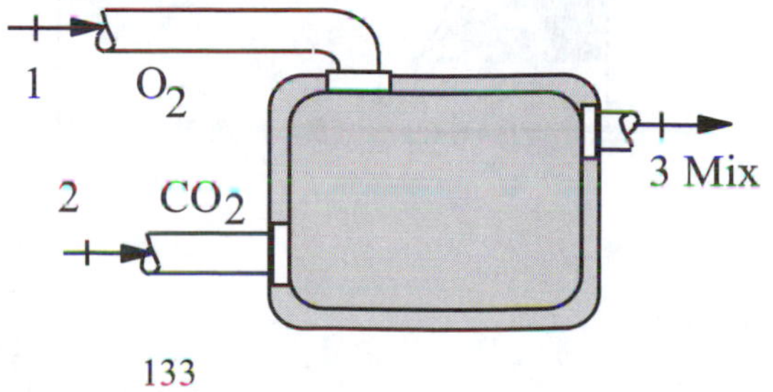

12.3 A mixture used in a refrigerator

A compressor in a refrigerator uses a new refrigerant R-410, which is 50% R-32 and 50% R-125 by mass. The inlet state is –10°C, 200 kPa and it is compressed up to 1000 kPa.

a) How high is the compressor exit temperature assuming reversible process?
b) How high is it if we assume a compressor isentropic efficiency of 85%.

Solution:

C.V. Ideal Compressor

Process: $q = 0$; adiabatic and reversible.

Energy Eq.6.13: $w = h_i - h_e$;

Entropy Eq.9.8: $s_e = s_i + s_{gen} + \int dq/T = s_i + 0 + 0 = s_i$

From Eq.12.15:

$$R_{mix} = \sum c_i R_i = \frac{1}{2} \times 0.1598 + \frac{1}{2} \times 0.06927 = 0.1145 \text{ kJ/kg K}$$

FromEq.12.23:

$$C_{P\,mix} = \frac{1}{2} \times 0.822 + \frac{1}{2} \times 0.791 = 0.8065 \text{ kJ/kg K}$$

$$R_{mix}/\ C_{P\,mix} = 0.1145/0.8065 = 0.14197$$

For constant s, ideal gas and use constant specific heat as in Eq.8.29

$$T_e/T_i = (P_e/P_i)^{R/Cp}$$

$$T_e = 263.15 \times (1000/200)^{0.14197} = \mathbf{330.7\ K}$$

$$w_{c\,s} \cong C_{P\,mix}(\ T_i - T_{e\,s}) = 0.8065\ (263.15 - 330.7) = \mathbf{-54.48\ kJ/kg}$$

Now do the actual compressor.

$$w_{c\,ac} = w_{c\,s}/\eta_c \cong C_{P\,mix}(\ T_i - T_{e\,ac}) = C_{P\,mix}(1/\eta_c)(T_i - T_{e\,s})$$

$$T_{e\,ac} = T_i - (1/\eta_c)(T_i - T_{e\,s}) = 263.15 - (263.15 - 330.7)/0.85$$

$$= \mathbf{342.6\ K = 69.5^oC}$$

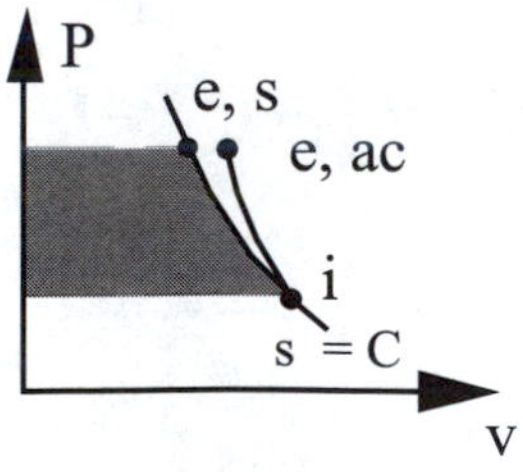

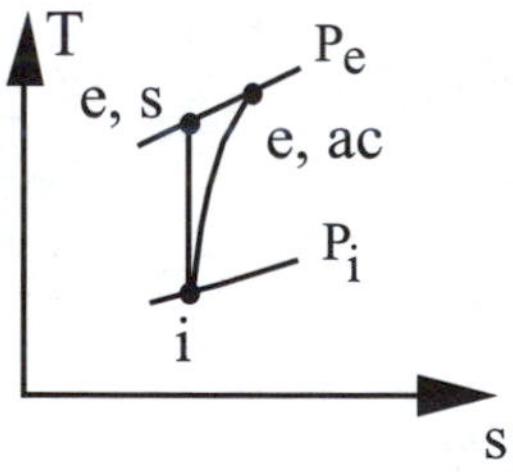

12.4 A mixing and cooling process

A flow of 0.1 kg/s carbon dioxide at 2000 K, 150 kPa is mixed with a flow of 0.4 kg/s water at 400 K, 150 kPa and after the mixing it goes through a heat exchanger being heated by a 1200 K reservoir. The resulting exit mixture is at 1000 K and 125 kPa (i.e. there was a slight pressure loss). Use table A.8 to solve this problem. How much heat transfer is needed? What is the entropy generation rate for the whole process?

Solution:

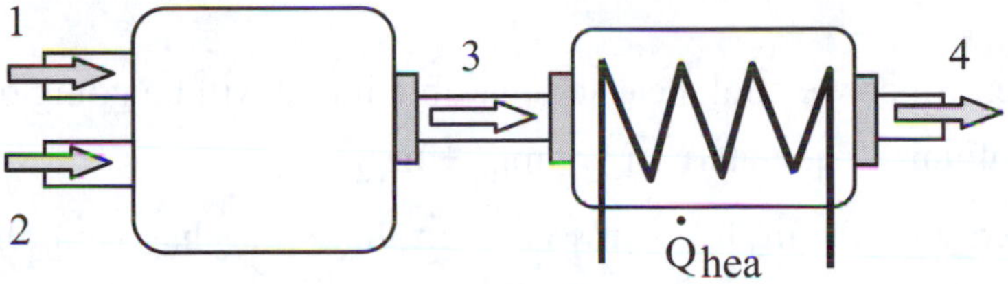

C.V. Total mixing section and heat exchanger. Steady flow and no work. To do the entropy at the partial pressures we need the mole fractions.

$$\dot{n}_{H2O} = \dot{m}_{H2O}/M_{H2O} = 0.4 / 18.015 = 0.0222 \text{ kmol/s}$$

$$\dot{n}_{CO2} = \dot{m}_{CO2}/M_{CO2} = 0.1 / 44.01 = 0.0022722 \text{ kmol/s}$$

$$y_{CO2} = \frac{0.0022722}{0.0222 + 0.0022722} = 0.09283, \quad y_{H2O} = 1 - y_{CO2} = 0.90717$$

Energy Eq.: $\dot{m}_{CO2}\, h_1 + \dot{m}_{H2O}\, h_2 + \dot{Q} = \dot{m}_{H2O}\, h_{4\,H2O} + \dot{m}_{CO2}\, h_{4\,CO2}$

Entropy Eq.: $\dot{m}_{CO2}\, s_1 + \dot{m}_{H2O}\, s_2 + \dfrac{\dot{Q}}{T_{source}} + \dot{S}_{gen} = \dot{m}_{H2O}\, s_{4\,H2O} + \dot{m}_{CO2}\, s_{4\,CO2}$

We use Table A.8 for properties on a mass basis and Eq.8.28 for change in entropy.

	1 CO2	2 H2O	4 H2O	4 CO2
h [kJ/kg]	2290.51	742.4	1994.13	971.67
s^o_T [kJ/kg K]	7.0278	11.0345	12.9192	6.119

$$\dot{Q} = \dot{m}_{H2O}\,(h_{4\,H2O} - h_2) + \dot{m}_{CO2}\,(h_{4\,CO2} - h_1)$$

$$= 0.4\,(1994.13 - 742.4) + 0.1\,(971.67 - 2290.51) = \mathbf{368.8\ kW}$$

$$\dot{S}_{gen} = \dot{m}_{H2O}\,(s_{4\,H2O} - s_2) + \dot{m}_{CO2}\,(s_{4\,CO2} - s_1) - \frac{\dot{Q}}{T_{source}}$$

$$= 0.4\,[12.9192 - 11.0345 - 0.4615\,\ln(0.90717 \times \frac{125}{150})\,]$$

$$+\, 0.1\,[6.119 - 7.0278 - 0.1889\,\ln(0.09283 \times \frac{125}{150})\,] - \frac{368.8}{1200}$$

$$= -0.042535 + 0.80552 - 0.30733 = \mathbf{0.456\ kW/K}$$

12.5 A moist air cooling process

A flow of 1 m^3/s moist air at 40°C and 50% relative humidity is brought through an air conditioner unit where it is cooled to 17°C and returned. How much heat transfer is needed and how much water is condensed out if any?

Solution:

C.V. Cooler. Since we cool, heat and possible liquid will be going out.

Continuity Eq. water: $\dot{m}_{v1} = \dot{m}_{liq} + \dot{m}_{v2}$

Energy Eq.: $\dot{m}_a h_{a1} + \dot{m}_{v1} h_{v1} = \dot{m}_a h_{a2} + \dot{m}_{v2} h_{v2} + \dot{m}_{liq} h_f + \dot{Q}_{out}$

Let us find the air flow rate so we know the scaling from state 1

B.1.1: $P_{g1} = 7.384$ kPa , $P_{v1} = \phi_1 P_{g1} = 3.692$ kPa

$\dot{m}_a = P_a\dot{V}/R_aT = (100 - 3.692)\ 1\ /(0.287 \times 313.15) = 1.072$ kg/s

Solve the problem using the **tables and formulas**:

State 1: $P_g (T_{dew}) = P_{v1}$ => $T_{dew} = 27.4$°C since $T_2 = 17$°C we have condensation

$\omega_1 = 0.622\ P_{v1}\ /(P - P_{v1}) = 0.622 \times 3.692\ /\ (100 - 3.692) = 0.0238$

State 2: $P_{v2} = P_{g2} = 1.959$ kPa => $\omega_2 = 0.622 \times 1.959\ /\ 98.04 = 0.01243$

B.1.1: $h_{v1} = 2574.3$ kJ/kg, $h_{v2} = 2532.6$ kJ/kg, $h_f = 71.4$ kJ/kg

Flow of liquid out becomes

$$\dot{m}_{liq} = \dot{m}_a (\omega_1 - \omega_2) = 1.072\ (0.0238 - 0.01243) = \mathbf{0.0122\ kg/s}$$

The heat transfer becomes

$$\dot{Q}_{out} = \dot{m}_a(h_{a1} - h_{a2}) + \dot{m}_a\omega_1 h_{v1} - \dot{m}_a\omega_2\, h_{v2} - \dot{m}_{liq}\, h_f$$

$$= \dot{m}_a[C_P(T_1 - T_2) + \omega_1 h_{v1} - \omega_2\, h_{v2}] - \dot{m}_{liq}\, h_f$$

$$= 1.072[1.004(40\text{–}17) + 0.0238 \times 2574.3 - 0.01243 \times 2532.6] - 0.0122 \times 71.4$$

$$= \mathbf{55.8\ kW}$$

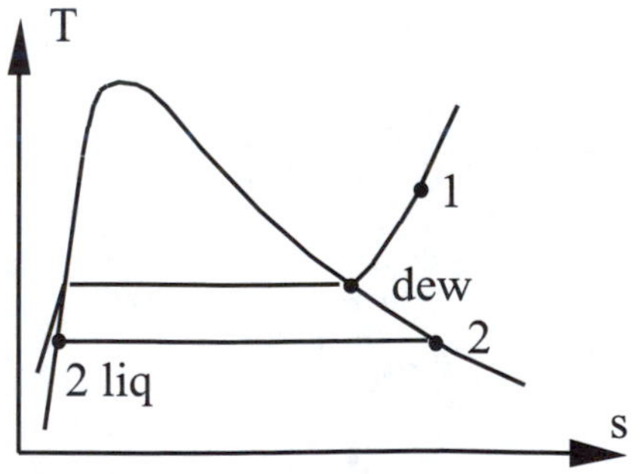

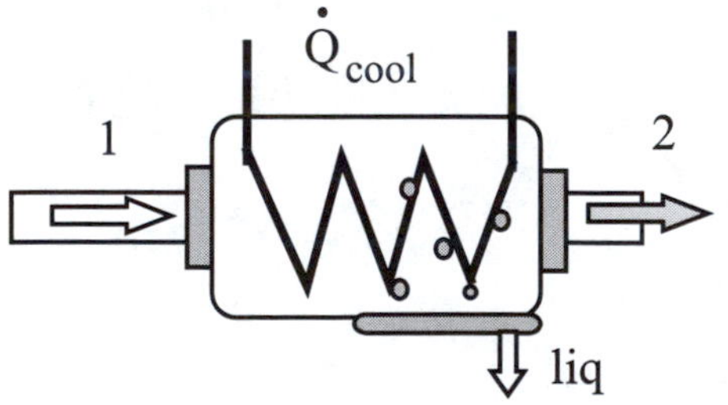

12.6 A moist air cooling process

A flow of 1 m^3/s moist air at 40°C and 50% relative humidity is brought through an air conditioner unit where it is cooled to 17°C and returned. How much heat transfer is needed and how much water is condensed out if any?

Solution:

C.V. Cooler. Since we cool, heat and possible liquid will be going out.

Continuity Eq. water: $\dot{m}_{v1} = \dot{m}_{liq} + \dot{m}_{v2}$

Energy Eq.: $\dot{m}_a h_{a1} + \dot{m}_{v1} h_{v1} = \dot{m}_a h_{a2} + \dot{m}_{v2} h_{v2} + \dot{m}_{liq} h_f + \dot{Q}_{out}$

Let us find the air flow rate so we know the scaling from state 1

B.1.1: $P_{g1} = 7.384$ kPa , $P_{v1} = \phi_1 P_{g1} = 3.692$ kPa

$\dot{m}_a = P_a\dot{V}/R_aT = (100 - 3.692)\ 1 /(0.287 \times 313.15) = 1.072$ kg/s

Solve the problem using the **psychrometric chart.**

Look up state 1 on the chart, page 735.

State 1: $\dot{m}_{v1}/\dot{m}_a = \omega_1 = 0.0232$, $\tilde{h}_1 = 118.3$ kJ/kg air, $T_{dew} = 27.5°C$

State 2: $T_2 < T_{dew} = 27.5°C \Rightarrow \phi_2 = 100\%$,

$\dot{m}_{v2}/\dot{m}_a = \omega_2 = 0.0122$, $\tilde{h}_2 = 68$ kJ/kg air

Liquid state at the lowest T from B.1.1: $h_f = 71.4$ kJ/kg

$\dot{m}_{liq} = \dot{m}_a\,(\omega_1 - \omega_2) = 1.072 \times (0.0232 - 0.0122) = \mathbf{0.01179\ kg/s}$

$\dot{Q}_{out} = \dot{m}_a\tilde{h}_1 - \dot{m}_{liq} h_f - \dot{m}_a \tilde{h}_2 = \dot{m}_a\,[\,\tilde{h}_1 - (\omega_1 - \omega_2)\,h_f - \tilde{h}_2]$

$= 1.072$ kg/s $[\,118.3 - (0.0232 - 0.0122) \times 71.4 - 68\,]$ kJ/kg

$= 1.072 \times 49.5 =$ **53 kW**

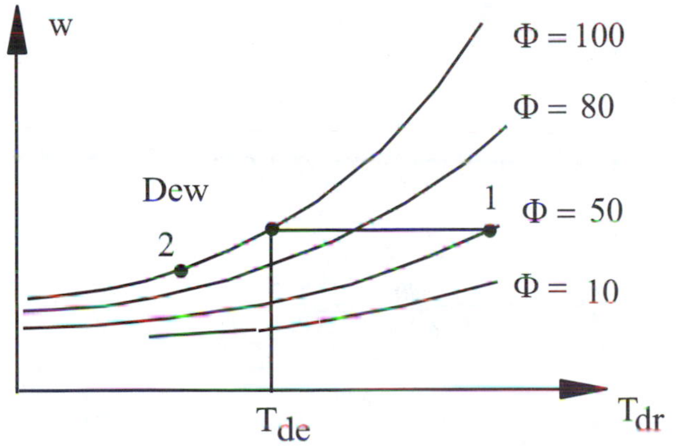

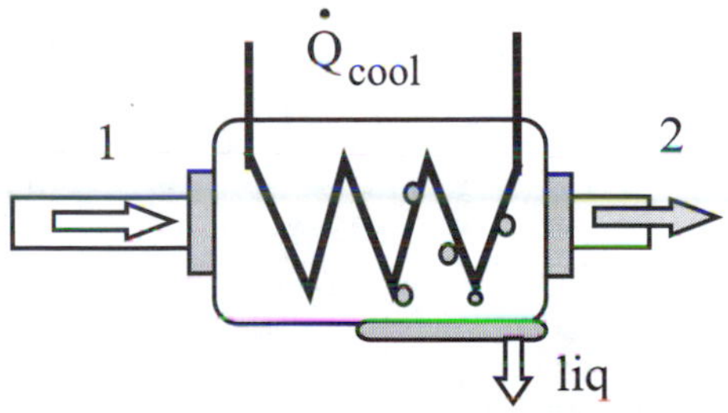

CHAPTER 13
STUDY PROBLEMS
THERMODYNAMIC RELATIONS

- **Clayperon equation**
- **Thermodynamic relations for h, u, s and compressibility**
- **Equations of state**
- **Compressibility factor and generalized charts**
- **Mixture property models**

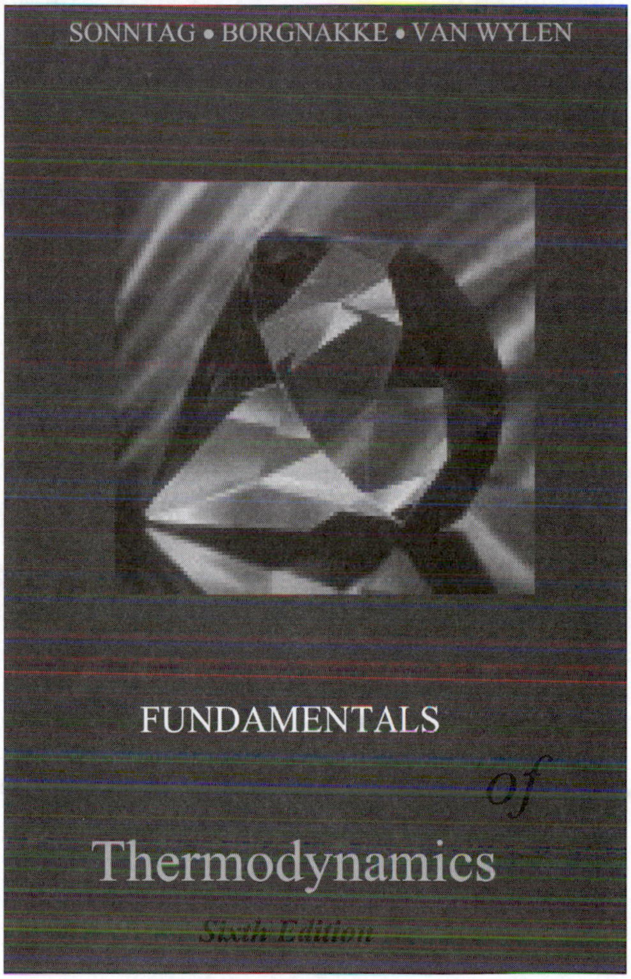

13.1 The Clapeyron equation

The saturation temperature for oxygen at room pressure, 101.3 kPa, is 90.2 K and it is desired to know the saturation pressure at a temperature of 75 K. The enthalpy of evaporation in that range is approximately equal to 220 kJ/kg.

Solution:

To answer the question we assume the vaporization line continues down to the 75 K before reaching the triple point. From Table 3.2 we know the triple point is at 54 K and 0.15 kPa. The oxygen table is not printed in the book, but it is included in the software version of the tables. We apply Clapeyron equation for the vaporization curve to get its mathematical form. This is shown in Eq.13.7 and integrated on page 514,

$$\frac{dP_{sat}}{dT} = \frac{h_{fg}}{Tv_{fg}} \approx \frac{h_{fg}P_{sat}}{RT^2} \Rightarrow \ln\frac{P_2}{P_1} = \frac{h_{fg}}{R}\left[\frac{1}{T_1} - \frac{1}{T_2}\right]$$

For $T_2 = 75$ K:

$$\ln\frac{P_2}{101.3} = \frac{220}{0.2598}\left[\frac{1}{90.2} - \frac{1}{75}\right] \quad => \quad P_2 = \mathbf{15.1\ kPa}$$

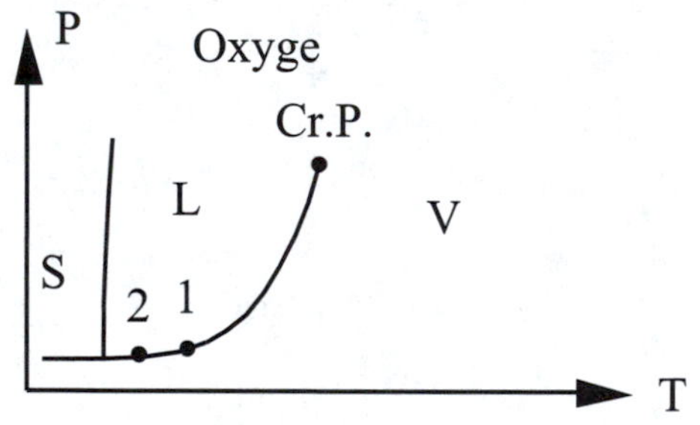

Comment: If you check with the software you will find a pressure of 14.5 kPa.

13.2 Equations of state applied to methane

Determine the pressure of 8.35 kg methane located in a 0.1 m^3 tank at 250 K. Compare the tables to the ideal gas model, van der Waal EOS and the Redlich-Kwong EOS

Solution:

The state is given by $(T = 250\ K,\ v = V/m = 0.1/8.35 = 0.011976\ m^3/kg)$

Table A.2 or B.7.2:

$T_c = 190.4$ K,

$P_c = 4600$ kPa

$$T_r = \frac{T}{T_c} = \frac{250}{190.4} = 1.313$$

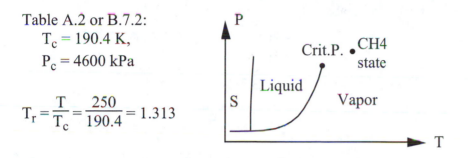

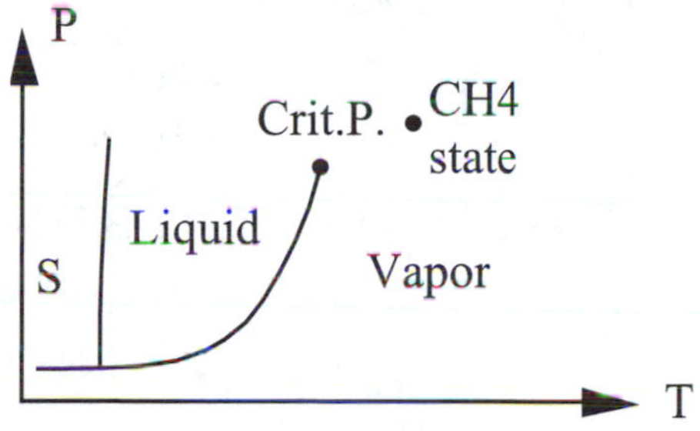

Locating the state in **Table B.7.2:** P = 8000 kPa, very close

Ideal gas model: $$P = \frac{RT}{v} = \frac{0.5183 \times 250}{0.011976} = 10\ 820\ \text{kPa}$$

For **van der Waal equation of state** from Table D.1 we have

$$b = \frac{1}{8}\frac{RT_c}{P_c} = 0.125 \times \frac{0.5183 \times 190.4}{4600} = 0.002\ 681\ 64\ m^3/kg,$$

$$a = 27\ b^2\ P_c = 27 \times (0.002\ 681\ 64)^2 \times 4600 = 0.893\ 15\ \text{kPa}\ (m^3/kg)^2$$

The equation is: $$P = \frac{RT}{v-b} - \frac{a}{v^2} = \frac{0.5183 \times 250}{0.011976 - 0.002\ 681\ 64} - \frac{0.89315}{0.011976^2} = 7714\ \text{kPa}$$

For **Redlich-Kwong equation of state** we have the parameters from Table D.1

$$b = 0.08664\frac{RT_c}{P_c} = 0.08664 \times \frac{0.5183 \times 190.4}{4600} = 0.001\ 858\ 7\ m^3/kg,$$

$$a = 0.42748\ T_r^{-1/2}\frac{R^2T_c^2}{P_c} = 0.42748 \times \frac{0.5183^2 \times 190.4^2}{1.313^{1/2} \times 4600} = 0.789809\ \text{kPa}\ (m^3/kg)^2$$

The equation is:

$$P = \frac{RT}{v-b} - \frac{a}{v^2 + bv} = \frac{0.5183 \times 250}{0.011976 - 0.0018587} - \frac{0.789809}{0.011976^2 + 0.0018587 \times 0.011976}$$

$$= 8040\ \text{kPa}$$

13.3 Speed of sound

Determine the speed of sound for ammonia at 400 kPa, 50°C. Use the printed tables and form the derivatives numerically.

Solution:

The definition of the speed of sound leads to the expression in Eq.13.43

$$c^2 = \left(\frac{\partial P}{\partial \rho}\right)_s = -v^2\left(\frac{\partial P}{\partial v}\right)_s$$

Superheated vapor water at 50°C, 400 kPa in Table B.2.2

$$v = 0.38226 \text{ m}^3\text{/kg}, \quad s = 5.7850 \text{ kJ/kg K}$$

At P = 500 kPa & s = 5.7850 kJ/kg K: T = 66.84°C, $v = 0.32141 \text{ m}^3\text{/kg}$

We can now form the derivative at constant s between the two pressure values

$$c^2 = -(0.38226)^2 \frac{\text{m}^6}{\text{kg}^2} \left(\frac{500 - 400}{0.32141 - 0.38226}\right) \frac{\text{kPa}}{\text{m}^3\text{/kg}}$$

$$= -0.146123 \times (-1643.385) \times 10^3 \text{ m}^2\text{/s}^2 = 2.40136 \times 10^5 \text{ m}^2\text{/s}^2$$

$$\Rightarrow \qquad c = \mathbf{490\ m/s}$$

13.4 Equations of state applied to methane

Redo Study problem 13.2 but use the generalized charts.

Determine the pressure of 8.35 kg methane located in a 0.1 m^3 tank at 250 K. Compare the tables to the ideal gas model, van der Waal EOS and the Redlich-Kwong EOS

Solution:

The state is given by (T = 250 K, v = V/m = 0.1/8.35 = 0.011976 m^3/kg)

Table A.2 or B.7.2: $T_c = 190.4$ K, $P_c = 4600$ kPa => $T_r = \frac{T}{T_c} = \frac{250}{190.4} = 1.313$

For the **generalized charts** we have the state as (T, v) so since that chart has an entry with (T_r, P_r) it becomes trial and error.

Guess: $P_r = 1.5$ & $T_r = 1.313$ => Chart Fig. D1, page 728: Z = 0.75

$$\Rightarrow \quad v = ZRT/P = 0.75 \times \frac{0.5183 \times 250}{1.5 \times 4600} = 0.01408 \text{ m}^3\text{/kg} \quad \text{(too large)}$$

Guess: $P_r = 2$ & $T_r = 1.313$ => Chart Fig. D1, page 728: Z = 0.70

$$\Rightarrow \quad v = ZRT/P = 0.70 \times \frac{0.5183 \times 250}{2 \times 4600} = 0.009859 \text{ m}^3\text{/kg} \quad \text{(too small)}$$

Interpolate: $P_r = 2 - 0.5 \frac{0.011976 - 0.009859}{0.01408 - 0.009859} = 1.75$

Check: $P_r = 1.75$ & $T_r = 1.313$ => Chart Fig. D1, page 728: Z = 0.73

$$\Rightarrow \quad v = ZRT/P = 0.73 \times \frac{0.5183 \times 250}{1.75 \times 4600} = 0.01175 \text{ m}^3\text{/kg} \quad \text{(close)}$$

This is probably as close as we can do with the error in reading the chart. Of course we could use the computer version of it and be more accurate.

$$P = P_r\, P_c = 1.75 \times 4600 = 8050 \text{ kPa}$$

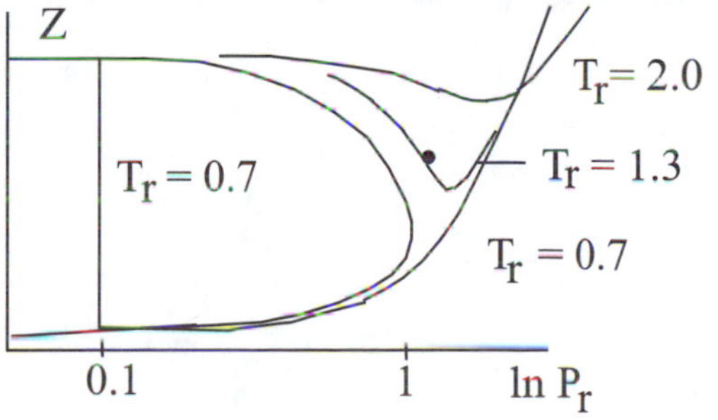

13.5 Generalized charts and enthalpy corrections

A piston cylinder contains 1 kg of propane at 296 K, 2 MPa which now expands in a constant pressure process to reach a temperature of 440 K. Find the best estimate for the needed work and heat transfer.

Solution:

The computer software does have the properties for propane but we will use the generalized charts to solve for this instead.

Table A.2: $T_c = 370$ K, $P_c = 4260$ kPa

$$=> \quad T_{r1} = \frac{T}{T_c} = \frac{296}{370} = 0.8, \quad P_{r1} = \frac{P}{P_c} = \frac{2000}{4260} = 0.47$$

$$=> \quad T_{r2} = \frac{T}{T_c} = \frac{440}{370} = 1.19, \; P_{r2} = \frac{P}{P_c} = \frac{2000}{4260} = 0.47$$

Table A.5: $R = 0.1886$ kJ/kg K, $C_p = 1.679$ kJ/kg K

From Fig. D.1 and D.2, page 728-729: $Z_1 = 0.07, \quad \Delta h_1 = 4.5$

From Fig. D.1 and D.2, page 728-729: $Z_2 = 0.89, \quad \Delta h_2 = 0.4$

$$v_1 = \frac{Z_1 R T_1}{P_1} = \frac{0.07 \times 0.1886 \times 296}{2000} = 0.001954 \text{ m}^3/\text{kg}$$

$$v_2 = \frac{Z_2 R T_2}{P_2} = \frac{0.89 \times 0.1886 \times 440}{2000} = 0.03693 \text{ m}^3/\text{kg}$$

$$h_2 - h_1 = C_p (T_2 - T_1) + RT_c (\Delta h_1 - \Delta h_2)$$

$$= 1.679(440 - 296) + 0.1886 \times 370\,(4.5 - 0.4) = 527.88 \text{ kJ/kg}$$

Process: $P = \text{constant} \quad => {}_1W_2 = mP_1 (v_2 - v_1)$

${}_1W_2 = 1 \times 2000\,(0.03693 - 0.001954) =$ **69.95 kJ**

Energy Eq.: ${}_1Q_2 = m(u_2 - u_1) + {}_1W_2 = m(h_2 - h_1) =$ **527.9 kJ**

13.6 Generalized charts and entropy corrections

Butane used in a Rankine cycle enters a reversible adiabatic turbine at 152°C, 5.7 MPa and it leaves the condenser as saturated liquid at 25°C. We want to determine if the turbine exit state is in the saturated two-phase region or in the superheated vapor region.

Solution:

The basic analysis of the turbine gives constant s through the turbine. The exit pressure is the saturation pressure for the 25°C in the condenser so the state is $(P, s = s_1)$.

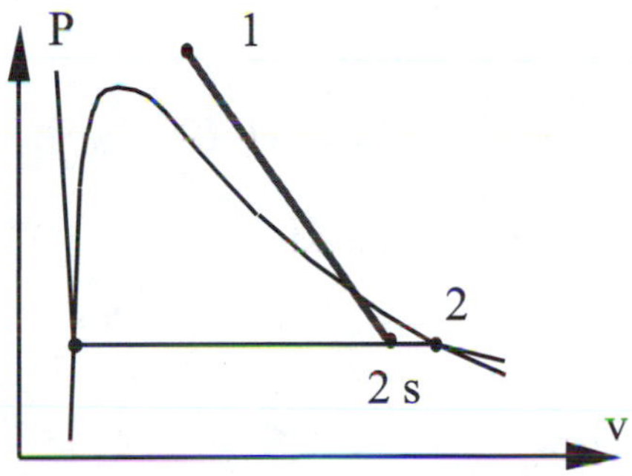

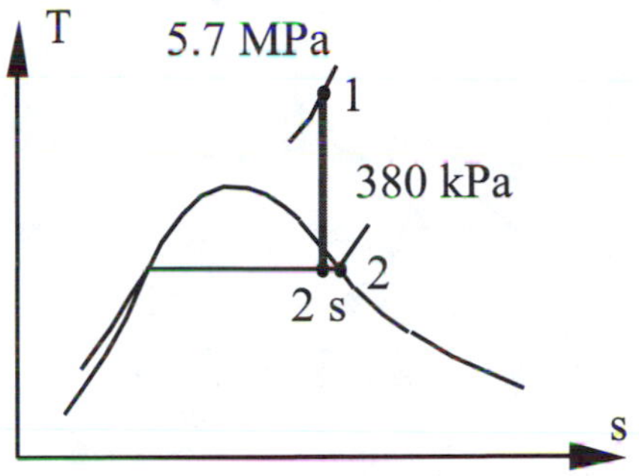

To see if the exit state is two-phase we compute $s_2 - s_{2'}$.

Table A.2: $T_c = 425.2$ K, $P_c = 3800$ kPa

$$\Rightarrow \quad T_{r1} = \frac{T}{T_c} = \frac{425.15}{425.2} = 1.0, \qquad P_{r1} = \frac{P}{P_c} = \frac{5700}{3800} = 1.5$$

$$\Rightarrow \quad T_{r2'} = \frac{T}{T_c} = \frac{298}{370} = 0.70, \qquad P_{r2'} = P_{r\ \text{sat at}\ T2'} = 0.1$$

Table A.5: $R = 0.1430$ kJ/kg K, $C_p = 1.716$ kJ/kg K

From Fig. D.3, page 730: $\Delta s_1 = 2.86$, $\Delta s_2 = 0.21$

$$s_2 - s_{2'} = s_1 - s_{2'} = C_p \ln(T_1/T_{2'}) - R \ln(P_1/P_{2'}) + R(\Delta s_2 - \Delta s_1)$$

$$= 1.716 \ln\left(\frac{1.0}{0.7}\right) - 0.143 \ln\left(\frac{1.5}{0.1}\right) + 0.143\ (0.21 - 2.86)$$

$$= 0.6121 - 0.3873 - 0.379 = -0.15 \text{ kJ/kg K} < 0$$

Notice here the T and P ratio's are both equal to the corresponding relative T and P ratios. Since now s_2 is smaller than the saturated vapor value we have **two-phase exit state**.

CHAPTER 14
STUDY PROBLEMS

COMBUSTION

- **Introduction and definitions**
- **Combustion of hydrocarbon fuels**
- **The energy equation**
- **The entropy equation**

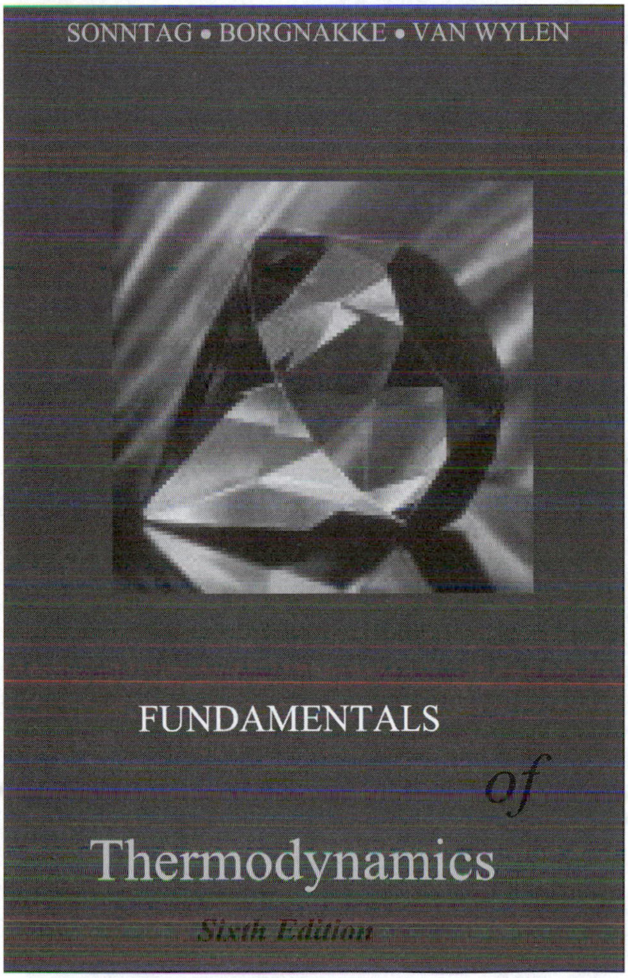

14.1 A combustion equation for JP-8 jet fuel

A modern jet engine or gas turbine uses JP-8 jet fuel, a mixture with the average composition as $C_{13}H_{23.8}$ and a molecular weight of 180.13. The combustion takes place with 125% theoretical air. What is the air-fuel ratio on a mole and mass basis (A/F) and what is the dew point temperature of the products at 100 kPa?

Solution:

The stoichiometric reaction equation is

$$C_{13}H_{23.8} + \nu_{O2}\left(O_2 + 3.76\,N_2\right) \rightarrow \nu_{CO2}\,CO_2 + \nu_{H2O}\,H_2O + \nu_{N2}\,N_2$$

Where the stoichiometric coefficients for CO_2 and H_2O comes from the fuel composition.

Continuity C: $13 = \nu_{CO2}$;

Continuity H: $23.8 = 2\,\nu_{H2O} \quad \Rightarrow \quad \nu_{H2O} = 11.9$

The **stoichiometric** (100% theoretical air) oxygen needed thus becomes

Continuity O: $2\nu_{O2} = 13 \times 2 + 11.9 = 37.9 \quad \Rightarrow \quad \nu_{O2} = 18.95;$

Then the **actual** combustion equation now has oxygen as

$$\nu_{O2\ ac} = 1.25 \times \nu_{O2\ stoich.} = 1.25 \times 18.95 = 23.6875$$

and nitrogen as

$$\nu_{N2} = 3.76\ \nu_{O2} = 3.76 \times 23.6875 = 89.065$$

Now we see the excess oxygen (23.6875 – 18.95 = 4.7375) in the products as

$$C_{13}H_{23.8} + 1.25 \times 18.95\left(O_2 + 3.76\,N_2\right) \rightarrow$$
$$13\,CO_2 + 11.9\,H_2O + 4.7375\,O_2 + 89.065\,N_2$$

The A/F ratio on a mole basis is

$$A/F = 23.6875 \times (1 + 3.76) / 1 = \mathbf{112.75}$$

The A/F ratio on a mass basis is

$$A/F = 23.6875 \times (31.999 + 3.76 \times 28.013) / 180.13 = \mathbf{18.06}$$

To find the product dew point we need to find the water partial pressure which is the mole fraction time the total pressure (assuming ideal gas mixture).

$$y_{H2O} = \frac{11.9}{13 + 11.9 + 4.7375 + 89.065} = 0.10$$

$$\Rightarrow \quad P_v = y_{H2O}\,P_{tot} = 0.1 \times 100\ \text{kPa} = 10\ \text{kPa} = P_{g\ Tdew}$$

Now look in Table B.1.2: $\mathbf{T_{dew} = 45.8^oC}$

14.2 Combustion of sulfur

Most coals have some amount of sulfur in it. Consider as a part of a combustion process that S is oxidized to SO_2 with stoichiometric air. The reactants are supplied at the reference T and P and the products after some heat transfer are at 1000 K. How much energy did the sulfur combustion provide per kmol of sulfur?

Solution:

Combustion equation: $S + \nu_{O2} (O_2 + 3.76\, N_2) \rightarrow 1\, SO_2 + \nu_{N2}\, N_2$

The O balance: $\nu_{O2} = 1$ => $\nu_{N2} = 3.76$

Now we can do the energy equation. Notice we do not have the SO_2 ideal gas tables so we use the heat capacity from Table A.5 (we could have used A.6 to be more accurate).

Energy Eq.: $H_R = H_R^{\circ} + \Delta H_R = H_R^{\circ} = H_P + Q = H_P^{\circ} + \Delta H_P + Q$

$$H_R^{\circ} = \bar{h}^0_{f\,S} + \bar{h}^0_{f\,O2} + 3.76\,\bar{h}^0_{f\,N2} = 0 + 0 + 0 = 0 \quad \text{(S from A.10)}$$

$$H_P^{\circ} = \bar{h}^0_{f\,SO2} + 3.76\,\bar{h}^0_{f\,N2} = -296\,842 + 0 = -296\,842 \text{ kJ/kmol}$$

$$\Delta H_P = C_p\, M\, (T_P - T_o) + 3.76\ \Delta\bar{h}_{N2}$$

$$= 0.624 \times 64.059\, (1000 - 298.15) + 3.76 \times 21\,463$$

$$= 28\,055 + 80\,701 = 108\,756 \text{ kJ/kmol}$$

Now solve for Q

$$Q = H_R^{\circ} - H_P^{\circ} - \Delta H_P$$

$$= 0 - (-296\,842) - 108\,756 = \mathbf{188\,086\ kJ/kmol}$$

Comment: Today we do what we can to take the sulfur out of the fuel (oil or coal) before we burn it. Water in the products together with sulfur dioxide will generate sulfuric acid causing acid rain. This previously caused sections of forrest to die out.

14.3 Heat release by JP-8 jet fuel

A modern jet engine or gas turbine uses JP-8 jet fuel, see Table 14.3. It has a molecular weight of 180.13. The combustion takes place with 125% theoretical air and the fuel is added as a liquid. How much air per kg fuel does it require and what is the specific energy released by the combustion process ($q = -H_{RP}$ per kg mixture). If the energy is used as the heat input in a gasturbine with cycle efficiency of 45% how much fuel (kg/s) is needed to have a power output of 10 MW?

Solution:

The stoichiometric reaction equation is

$$C_{13}H_{23.8} + \nu_{O2}\left(O_2 + 3.76\, N_2\right) \rightarrow 13\, CO_2 + 11.9\, H_2O + \nu_{N2}\, N_2$$

Where the stoichiometric coefficients for CO_2 and H_2O comes from the fuel composition as shown in Study problem 14.1. The oxygen needed becomes

$$\nu_{O2} = 13 + 11.9/2 = 18.95; \qquad \nu_{N2} = 3.76\, \nu_{O2} = 71.252$$

The actual combustion equation now has excess oxygen as

$$C_{13}H_{23.8} + 1.25 \times 18.95\left(O_2 + 3.76\, N_2\right) \rightarrow$$

$$13\, CO_2 + 11.9\, H_2O + 4.7375\, O_2 + 89.065\, N_2$$

The A/F ratio on a mass basis is

$$AF = 1.25 \times 18.95\ (31.999 + 3.76 \times 28.013) / 180.13 = \mathbf{18.06}$$

The enthalpy of combustion H_{RP} from Table 14.3 is –42 800 kJ/kg fuel taken for liquid fuel and water as vapor. In order to use that in a cycle calculation we need to convert it to energy per kg mixture.

$$m_{mix} = m_{fuel} + m_{air} = m_{fuel}\ (1 + AF)$$

We therefore have

$$q = -H_{RP} / (1 + AF) = 42\ 800 / 19.06 = \mathbf{2245\ kJ/kg}$$

For the cycle we then have

$$w_{net} = \eta\, q = 0.45 \times 2245 = 1010 \text{ kJ/kg}$$

$$\dot{m}_{mix} = \dot{W}/w_{net} = \frac{10\ 000}{1010} = 9.9 \text{ kg/s}$$

$$\dot{m}_{fuel} = \frac{\dot{m}_{mix}}{1 + AF} = \frac{9.9}{19.06} = \mathbf{0.52\ kg/s}$$

14.4 Combustion of n-Cetane

Consider a combustion process with n-Cetane $C_{16}H_{34}$, see Table 14.3, in a stoichiometric ratio with air. We notice that there is no entry for this fuel in Table A.10 yet we want to know how much energy we need to supply to a carburetor to vaporize the liquid fuel before we mix it with air. Secondly we also want to know the adiabatic flame temperature assuming the reactants are supplied at the reference P and T with the fuel as a vapor.

The stoichiometric reaction equation is

$$C_{16}H_{34} + \nu_{O2} (O_2 + 3.76\, N_2) \rightarrow 16\, CO_2 + 17\, H_2O + \nu_{N2}\, N_2$$

where we have applied the C and H atom balance already. The required oxygen becomes

$$\nu_{O2} = 16 + 17/2 = 24.5; \qquad \nu_{N2} = 3.76\, \nu_{O2} = 92.12$$

The energy equation becomes

Energy Eq.: $$H_R = H^o_R + \Delta H_R = H^o_R = H_P = H^o_P + \Delta H_P$$

Solve for ΔH_P

$$\Delta H_P = H^o_R - H^o_P = -H^o_{RP} = HV = \bar{h}^0_{f\,fuel} - 16\, \bar{h}^o_{f\,CO2} - 17\, \bar{h}^o_{f\,H2O}$$

From Table 14.3 we find for the liquid fuel and water as vapor $-H^o_{RP} = 44\,000$ kJ/kg. If we look at the fuel vapor and water as vapor we get $-H^o_{RP} = 44\,358$ kJ/kg so the difference is the fuel enthalpy of vaporization as the other terms cancel out

$$h_{fg} = HV_{fuel\ vap} - HV_{fuel\ liq} = h^o_{f\,fuel\ vap} - h^o_{f\,fuel\ liq} = 44\,538 - 44\,000 = \mathbf{538\ kJ/kg}$$

For the energy equation we need the HV per kmol of fuel so

$$HV = -H^o_{RP} = 44\,000\, \frac{kJ}{kg} = 44\,000 \times (16 \times 12.011 + 34 \times 1.008) = 9\,963\,712\, \frac{kJ}{kmol}$$

$$\Delta H_P = 16\, \Delta\bar{h}_{CO2} + 17\, \Delta\bar{h}_{H2O} + 92.12\, \Delta\bar{h}_{N2} = 9\,963\,712\ \text{kJ/kmol}$$

Now find the temperature for which the enthalpy terms, Table A.9, adds up. We start with the average [79 633 = 9 963 712 /(92.12 + 17 +16)] for nitrogen giving 2600+ K

2600K: $\Delta H_P = 16 \times 128074 + 17 \times 104520 + 92.12 \times 77963 = 11\,007\,975$ kJ/kmol **high**

2400K: $\Delta H_P = 16 \times 115779 + 17 \times 93\,741 + 92.12 \times 70\,640 = 9\,953\,418$ kJ/kmol **close**

Interpolate to get: $$T = 2400 + 200\, \frac{9\,963\,712 - 9\,953\,418}{11\,007\,975 - 9\,953\,418} = \mathbf{2402\ K}$$

14.5 Combustion of carbon monoxide

Assume a steady flow of 0.1 kg/s carbon monoxide CO and a stoichiometric amount of air flowing into a mixing chamber both at reference temperature and pressure. A complete combustion process generates carbon dioxide CO_2. What is the product exit temperature assuming an adiabatic process?

Solution:

The reaction equation is

$$CO + \nu_{O2}\left(O_2 + 3.76\, N_2\right) \rightarrow 1\ CO_2 + \nu_{N2}\, N_2$$

With the carbon balance ($\nu_C = 1$) done, so the oxygen balance becomes

$$0.5 + \nu_{O2} = 1 \qquad => \quad \nu_{O2} = 0.5, \qquad \nu_{N2} = 1.88$$

The energy equation for the whole setup has no Q or W terms so it is

Energy Eq.: $$H_R = H^o_R + \Delta H_R = H^o_R = H_P = H^o_P + \Delta H_P$$

Solve for ΔH_P

$$\Delta H_P = H^o_R - H^o_P = \bar{h}^o_{f\,CO} - \bar{h}^o_{f\,CO2}$$

$$= -110\ 527 - (-393\ 522) = 282\ 995 \text{ kJ/kmol CO}$$

Now the energy equation reads

$$\Delta H_P = \Delta\bar{h}_{CO2} + 1.88\ \Delta\bar{h}_{N2} = 282\ 995 \text{ kJ/kmol CO}$$

And we need to find the temperature for which the left hand side adds up to the proper value. The enthalpy terms are from Table A.9 so we have trial and error.

$$LHS_{2000} = \ 91\ 439 + 1.88 \times 56\ 137 = 196\ 976 \ \text{ too small}$$

$$LHS_{3000} = 152\ 853 + 1.88 \times 92\ 715 = 327\ 157 \ \text{ too large}$$

$$LHS_{2600} = 128\ 074 + 1.88 \times 77\ 963 = 274\ 644 \ \text{ too small}$$

$$LHS_{2800} = 140\ 435 + 1.88 \times 85\ 323 = 300\ 842 \ \text{ too large}$$

Now interpolate

$$T = 2600 + 200 \times \frac{282\ 995 - 274\ 644}{300\ 842 - 274\ 644} = \mathbf{2664\ K}$$

14.6 Entropy generation burning propene

Propene, C_3H_6, is burned with air in a steady flow burner. The reactants are supplied at the reference pressure and temperature and the mixture is lean so the adiabatic flame temperature is kept to 1800 K. What is the A/F ratio on a mole basis? What is the entropy generation per kmol fuel, neglecting all the partial pressure corrections?

Solution:

The reaction equation for a mixture with excess air is:

$$C_3H_6 + \nu_{O2}\left(O_2 + 3.76\, N_2\right) \rightarrow\ 3\, H_2O + 3\, CO_2 + 3.76\nu_{O2}\, N_2 + (\nu_{O2} - 4.5)O_2$$

Energy Eq.: $\quad H_R = H_R^\circ + \Delta H_R = H_R^\circ = H_P = H_P^\circ + \Delta H_P$

The entropy equation: $\quad S_R + S_{gen} = S_P \quad => \quad S_{gen} = S_P - S_R = S_P - S_R^\circ$

From table A.9 at reference T

$$\Delta H_R = \Delta h_{Fu} + \nu_{O2}(\Delta h_{O2} + 3.76\ \Delta h_{N2}) = 0$$

From table A.9 at 1800 K:

$$\Delta H_P = 3\ \Delta h_{H2O} + 3\ \Delta h_{CO2} + 3.76\ \nu_{O2}\ \Delta h_{N2} + (\nu_{O2} - 4.5)\ \Delta h_{O2}$$

$$= 3 \times 62\,693 + 3 \times 79432 + 3.76\ \nu_{O2} \times 48\,979 + (\nu_{O2} - 4.5)\ 51\,674$$

$$= 193\,842 + 235\,835\ \nu_{O2}$$

From table 14.3: $\quad H_P^\circ - H_R^\circ = H_{RP}^\circ = 42.081(-45\,780) = -1\,926\,468$ kJ/kmol

Now substitute all terms into the energy equation

$$-1\,926\,468 + 193\,842 + 235\,835\ \nu_{O2} = 0$$

Solve for ν_{O2}: $\quad \nu_{O2} = \dfrac{1\,926\,468 - 193\,842}{235\,835} = 7.3468, \quad \nu_{N2} = 27.624$

$A/F = 4.76\ \nu_{O2} / 1 =$ **34.97** $\quad$ [$(A/F) / (A/F)_S = 7.3468/4.5 = 1.633$]

Table A.9 contains the entropies at 100 kPa so we get:

$$S_P = 3 \times 259.452 + 3 \times 302.969 + (7.3468 - 4.5)\ 264.797 + 27.624 \times 248.304$$

$$= 9300.24 \text{ kJ/kmol fuel}$$

$$S_R = 267.066 + 7.3468 \times 205.148 + 27.624 \times 191.609 = 7067.25 \text{ kJ/kmol fuel}$$

$S_{gen} = 9300.24 - 7067.25 =$ **2233 kJ/kmol fuel**

14.7 Combustion of carbon monoxide

Assume a steady flow of carbon monoxide CO and a stoichiometric amount of air flowing into a mixing chamber both at reference temperature and pressure. A complete combustion process generates carbon dioxide CO_2. How much entropy is generated in the mixing chamber and how much is generated in the combustion process both per kmol CO?

Solution:

The reaction equation is

$$CO + \nu_{O2}\left(O_2 + 3.76\, N_2\right) \rightarrow 1\; CO_2 + \nu_{N2}\, N_2$$

so the oxygen balance becomes

$$0.5 + \nu_{O2} = 1 \qquad => \quad \nu_{O2} = 0.5, \qquad \nu_{N2} = 1.88$$

In the mixing process we will have the same temperature out and the total pressure out equals the inlet pressure. The entropy generation is only caused by the mixing process for which we need the molfractions and recall the air was mixed already.

$$y_{CO} = \frac{1}{1 + 0.5 + 1.88} = 0.2958; \qquad y_{air} = 1 - y_{CO} = 0.7042$$

Now the increase in entropy equals the generation due to the reduction to partial pressures

$$S_{gen\ mix} = (S_2 - S_1)_{CO} + (S_2 - S_1)_{air}$$

$$= 1(-\bar{R} \ln y_{CO}) + 4.76\,(-\bar{R} \ln y_{air})$$

$$= (-8.3145 \ln 0.2958) + 4.76\,(-8.3145 \ln 0.7042)$$

$$= \mathbf{24.0\ kJ/Kmol\ CO\ K}$$

The entropy equation for the whole setup has no Q term so it is

Entropy Eq.: $\qquad S_R + S_{gen\ tot} = S_P$

The entropy for the products are evaluated at the adiabatic flame temperature, see study problem 14.5 where we found $T = 2664$ K. For the products we have the molfractions

$$y_{CO2} = \frac{1}{1 + 1.88} = 0.3472\,; \qquad y_{N2} = 1 - y_{CO2} = 0.6528$$

$$S_P = S_{CO2} + 1.88\, S_{N2} = 326.773 - 8.3145 \ln(0.3472)$$

$$+ 1.88\left[262.488 - 8.3145 \ln(0.6528)\right] = 835.712 \text{ kJ/kmol K}$$

The CO comes in at reference P and the air is mixed

$$S_R = S_{CO} + 0.5\, S_{O2} + 1.88\, S_{N2} = 197.651 + 0.5[\,205.148 - 8.3145 \ln(0.21)\,]$$

$$+ 1.88\,[191.609 - 8.3145 \ln(0.79)\,] = 670.622 \text{ kJ/kmol K}$$

$$S_{gen\ tot} = S_P - S_R = 835.712 - 670.622 = 165.089 \text{ kJ/kmol K}$$

$$S_{gen\ comb} = S_{gen\ tot} - S_{gen\ mix} = 165.089 - 24.0 = \mathbf{141.1\ kJ/kmol\ CO\ K}$$

CHAPTER 15
STUDY PROBLEMS

CHEMICAL EQUILIBRIUM

- **General equilibrium considerations**
- **Chemical equilibrium**
- **Dilution and simultaneous reactions**
- **Ionization**

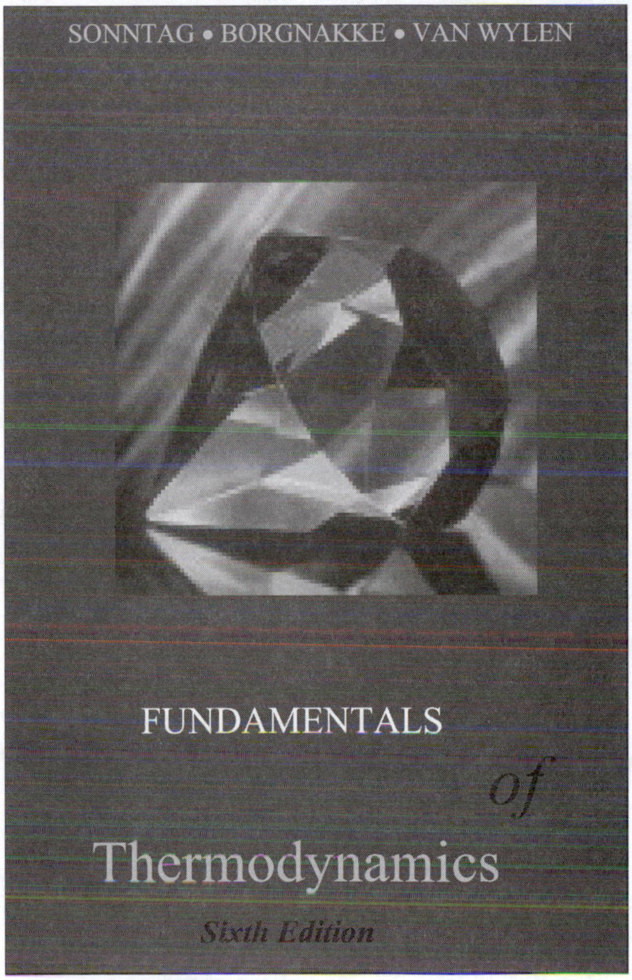

15.1 Dissociation of a diatomic gas

Assume we have a flow of a hydrogen gas that is heated up to 2800 K at constant pressure of 100 kPa. What is the percent (mole basis) of atomic hydrogen in the exit flow? To what temperature should it be heated to find 5% atomic hydrogen?

Solution:

The equilibrium reaction system and the equilibrium constant is:

At 2000 K, Table A.11

$\ln K = -5.005$

$K = 0.006704$

Reac.:	H_2 ⇔	$2\ H$
Initial	1	0
Change	-z	+2z
Equil.	1-z	2z

We have $P = P^o = 0.1$ MPa, and $n_{tot} = 1 + z$, so from Eq.15.16

$$K = \frac{y_H^2}{y_{H2}} \left(\frac{P}{P^0}\right)^{2-1} = \left(\frac{2z}{1+z}\right)^2 \left(\frac{1+z}{1-z}\right) (1) = 0.006704 ;$$

Reduce the powers in z to

$$\frac{4z^2}{1 - z^2} = 0.006704 \quad => \quad 4.006704\ z^2 = 0.006704 \quad => \quad z = 0.040905$$

Now the concentrations are

$$y_H = \frac{2z}{1+z} = \mathbf{0.0786} ; \qquad y_{H2} = \frac{1-z}{1+z} = 0.9214$$

If the concentration of atomic hydrogen should be 5% then we get

$$y_H = \frac{2z}{1+z} = 0.05 \quad => \quad z = 0.05/(2 - 0.05) = 0.025641$$

Then we find K as

$$K = \left(\frac{2z}{1+z}\right)^2 \left(\frac{1+z}{1-z}\right) (1) = 0.0026316 = \exp(-5.94017)$$

Now interpolate in Table A.10 to find

$$T = 2600 + 200 \frac{-5.94017 - (-6.519)}{-5.005 - (-6.519)} = \mathbf{2676\ K}$$

15.2 Heating of carbon dioxide

A constant pressure heater has a flow in of carbon dioxide at 300 K, 100 kPa and the exit flow is an equilibrium mixture of CO_2, CO and O_2 at 2800 K. Find the exit equilibrium mixture composition (mole fractions) and the needed heat transfer per kmol CO_2 in.

Solution:

The equilibrium reaction system and the equilibrium constant is:

At 2800 K, Table A.11

$\ln K = -3.781$

$K = 0.02279988$

Reac.:	$2\,CO_2 \Leftrightarrow$	$2\,CO$ +	$1\,O_2$
Initial	1	0	0
Change	-2z	+2z	+z
Equil.	1-2z	2z	z

We have $P = P^o = 0.1$ MPa, and $n_{tot} = 1 + z$, so from Eq.15.16

$$K = \frac{y_{CO}^2 y_{O_2}}{y_{CO_2}^2}\left(\frac{P}{P^0}\right) = \left(\frac{2z}{1-2z}\right)^2\left(\frac{z}{1+z}\right)(1) = 0.02279988\ ;$$

$$z^3 = (0.02279988 / 4)\,(1 - 2z)^2\,(1 + z)$$

Now evaluate the RHS for z = 0 (start), then repeatedly solve for a new z as follows:

$$z_{n+1}^3 = 0.00569997\,(1 - 2z_n)^2(1 + z_n)$$

This is called successive substitutions. Result: $z = 0.148$

$$y_{CO} = \frac{2z}{1+z} = \mathbf{0.258}, \quad y_{CO2} = \frac{1-2z}{1+z} = \mathbf{0.613}, \quad y_{O2} = \frac{z}{1+z} = \mathbf{0.129}$$

The energy equation (scaled to 1 kmol CO_2) becomes

$$Q = H_{ex} - H_{in} = y_{CO}\,\bar{h}_{CO} + y_{CO2}\,\bar{h}_{CO_2} + y_{O2}\,\bar{h}_{O_2} - \bar{h}_{CO_2\,in}$$

From Table A.9 we have the enthalpies (all in kJ/kmol):

$$\Delta\bar{h}_{CO} = 86\,070, \quad \bar{h}^0_{f\,CO} = -110\,527, \quad \Delta\bar{h}_{CO_2} = 140\,435, \quad \bar{h}^0_{f\,CO_2} = -393\,522$$

$$\Delta\bar{h}_{O_2} = 90\,080, \quad \Delta\bar{h}_{CO_2\,in} = 69$$

so

$$Q = 0.258(86\,070 - 110\,527) + 0.613(140\,435 - 393\,522) + 0.129(90\,080) - (69 - 393\,522)$$

$$= \mathbf{243\,621\ kJ/kmol}$$

Comment: If we assumed we heated carbon dioxide the heat transfer would be

$$Q = \Delta\bar{h}_{CO_2} = 140\,435 - 69 = 140\,366\ \text{kJ/kmol}$$

i.e. the formation of CO requires a significant amount of energy.

15.3 Dissociation of a diatomic gas with dilution

Let us look at the dissociation of diatomic hydrogen gas, but assume the hydrogen gas is diluted with Ar so the incoming concentration of hydrogen is 50% on a mole basis.

The equilibrium reaction system and the equilibrium constant is:

At 2000 K, Table A.11

$\ln K = -5.005$

$K = 0.006704$

Reac.:	$H_2 \Leftrightarrow$	2 H	Ar
Initial	1	0	1
Change	-z	+2z	0
Equil.	1-z	2z	1

We have $P = P^o = 0.1$ MPa, and $n_{tot} = 2 + z$, so from Eq.15.16

$$K = \frac{y_H^2}{y_{H2}} \left(\frac{P}{P^0}\right)^{2-1} = \left(\frac{2z}{2+z}\right)^2 \left(\frac{2+z}{1-z}\right) (1) = 0.006704\ ;$$

Reduce the equation in z to

$$(4 + C)\, z^2 + C\, z - 2\, C = 0, \qquad C = 0.006704$$

$$z = \frac{-C}{2(4+C)} \pm \frac{1}{2(4+C)} \sqrt{C^2 + 8C(4+C)}$$

$$= -0.0008366 \pm 0.12479 \sqrt{0.2149325} = 0.05702 \quad (\text{only } z > 0)$$

Now the concentrations are

$$y_H = \frac{2z}{2+z} = \mathbf{0.05544}\ ; \quad y_{H2} = \frac{1-z}{2+z} = \mathbf{0.45842}\ ; \quad y_{Ar} = \frac{1}{2+z} = \mathbf{0.48614}$$

Comment: Now we see relative more of the hydrogen molecules have dissociated. The dilution pushed the reaction towards the side with the more moles because the partial pressures were reduced. So the effect of the argon is not in the reaction but in the y's.

15.4 Simultaneous reactions

A mixture of 1 kmol carbon dioxide, 2 kmol carbon monoxide and 2 kmol of oxygen at 25C, 150 kPa is heated at constant pressure to 3000 K. Find the composition of the exit equilibrium mixture of CO_2, CO, O_2 and O gases.

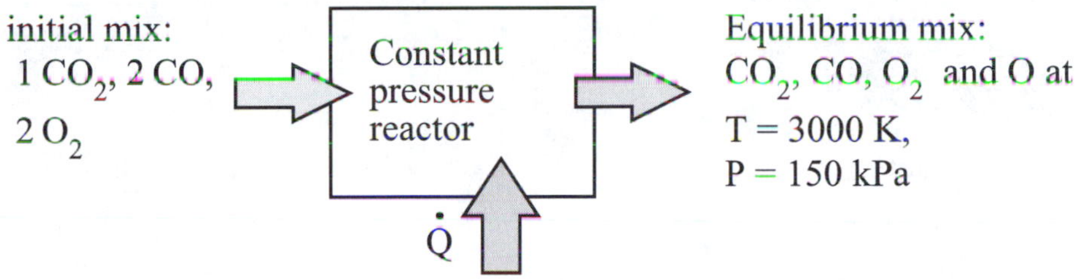

The carbon dioxide reaction

From A.10 at 3000 K:

$K_I = \exp(-2.217) = 0.108935$

Reaction	$2\,CO_2$	$\Leftrightarrow$	$2\,CO$	+	O_2
change	$-2x$		$+2x$		$+x$

The oxygen dissociation

From A.10 at 3000 K:

$K_{II} = \exp(-4.356) = 0.01283$

Reaction	O_2	$\Leftrightarrow$	$2\,O$
change	$-y$		$+2y$

Component:	CO_2	CO	O_2	O	
Initial	1	2	2	0	
Change	$-2x$	$2x$	$+x-y$	$+2y$	
Final	$1-2x$	$2+2x$	$2+x-y$	$2y$	$n = 5+x+y$

For each $n > 0 \quad \Rightarrow \quad -1 < x < +\frac{1}{2}\ ; \quad y > 0 \quad$ and $\quad 2 + x - y > 0$

Now write the two equilibrium constant equations, Eqs, 15.28-29, unknowns (x, y)

$$K_I = 0.108935 = \frac{y_{CO}^2\, y_{O2}}{y_{CO2}^2}\left(\frac{P}{P^0}\right)^1 = \left(\frac{2+2x}{1-2x}\right)^2\left(\frac{2+x-y}{5+x+y}\right)\left(\frac{150}{100}\right)$$

$$K_{II} = 0.01283 = \frac{y_O^2}{y_{O2}}\left(\frac{P}{P^0}\right)^1 = \left(\frac{2y}{5+x+y}\right)^2\left(\frac{5+x+y}{2+x-y}\right)\left(\frac{150}{100}\right)$$

Trial and error solution, $\quad x = -0.50476, \quad y = 0.1166$

$$y_{CO2} = \frac{1-2x}{5+x+y} = \mathbf{0.436}, \quad y_{CO} = \frac{2+2x}{5+x+y} = \mathbf{0.215},$$

$$y_{O2} = \frac{2+x-y}{5+x+y} = \mathbf{0.299}, \quad y_O = \frac{2y}{5+x+y} = \mathbf{0.05}$$

15.5 Ionization of nitrogen

Consider the ionization of nitrogen at 12 000 K, at which temperature we do not have molecular nitrogen N_2, but a mixture of N, N^+ and e^-. What should the pressure be to have a nitrogen ion concentration of 10%. The reaction constant for the ionization as $N \Leftrightarrow N^+ + e^-$ is $K = 0.0151$, see problem 15.78 page 647.

The reaction and the shifts

Reaction	N	$\Leftrightarrow$	N^+	+	e^-	
Initial	1		0		0	
change	$-x$		$+x$		$+x$	
Final	$1 - x$		$+x$		$+x$	$n = 1 + x$

$$\text{EQ:} \quad K = 0.0151 = \frac{y_{N+}y_{e-}}{y_N}\left(\frac{P}{P^0}\right) = \frac{\frac{x}{1+x}\frac{x}{1+x}}{\frac{1-x}{1+x}}\left(\frac{P}{P^0}\right)$$

Now: $\quad y_{N+} = 0.1 = \frac{x}{1+x} = y_{e-} \quad => \quad y_{N+} = 1 - 0.1 - 0.1 = 0.8$

And the equilibrium reaction equation becomes

$$K = 0.0151 = \frac{0.1 \times 0.1}{0.8}\left(\frac{P}{P^0}\right) = 0.0125\left(\frac{P}{P^0}\right)$$

$$P = \frac{0.0151}{0.0125} P^0 = \mathbf{120.8\ kPa}$$